Clean Water

Clean Water

*An Introduction to Water Quality
and Water Pollution Control*

~

Second Edition

Kenneth M. Vigil, P.E.
Environmental Engineer

Oregon State University Press
Corvallis

This is a thoroughly revised and updated edition of the author's 1996 book, *Clean Water: The Citizen's Complete Guide to Water Quality and Water Pollution Control* (Columbia Cascade Publishing).

The paper in this book meets the guidelines for permanence and durability of the Committee on Production Guidelines for Book Longevity of the Council on Library Resources and the minimum requirements of the American National Standard for Permanence of Paper for Printed Library Materials Z39.48-1984.

Library of Congress Cataloging-in-Publication Data
Vigil, Kenneth M.
  Clean water : an introduction to water quality and water pollution control /
Kenneth M. Vigil.— 2nd ed.
    p. cm.
Includes bibliographical references and index.
  ISBN 0-87071-498-8 (alk. paper)
  1. Water quality—Popular works. 2. Water—Purification—Popular works. I. Title.
  TD370 .V55 2003
  363.736'4--dc21

                                    2002151440

Oregon State University Press
500 Kerr Administration
Corvallis OR 97331
541-737-3166 • fax 541-737-3170
http://oregonstate.edu/dept/press

To my children—Luke, Maya, Serena—and your generation.
May you grow up in a world with an abundance
of cool, clear, clean, water.

Contents

Preface

Books about water quality and water pollution control are generally directed at an audience of engineers, scientists, lawyers, or other water quality professionals. Because of the intended audience, these books usually contain highly technical, detailed information. A single textbook may cover one topic only, such as a particular method of treating wastewater or the interpretation of a specific water quality regulation.

I wrote this book because I felt that information about water quality and water pollution control should be more accessible to students of environmental science and ecology in particular, and others interested in learning more about this field. I believed it was possible to summarize many of the important topics in a single book. I also felt it was possible to write the book in sufficient detail to be useful and informative without being so detailed that one must be a water quality professional to understand it. This book, then, is the result of my effort to create a complete introductory reference on water quality and water pollution control for use by students, educators, and the general public.

I wrote this book also because of my own interest in protecting the environment and my desire to help others learn more about clean water. I believe that informed citizens are more likely to make wise environmental decisions and encourage their political representatives to do likewise. I believe that working to maintain a clean environment is an important and worthwhile endeavor—not just for environmental professionals, but for all people.

Acknowledgments
I thank my family, friends, and colleagues for supporting and encouraging me while I was writing this book. Their encouragement provided me with extra energy when I needed it.

I also offer a general word of thanks to the many people who have helped me learn about water quality and water pollution control over the years. I am especially grateful to the faculty members of the Division of Environmental Engineering at Utah State University and staff from the Oregon Department of Environmental Quality. The foundation of my understanding of water quality was set while I was with both of these groups.

I am grateful to the staff at Oregon State University Press and Clarity Writing & Editing, Inc. for making this edition of the book a reality. I appreciated your ideas, suggestions, and professional support throughout the process.

Finally, I thank my wife, Roma, for understanding and encouraging me in all of my efforts.

Credits

All photos are by the author except as noted below.

Sam Lucero: Lake Superior and Lake Superior tributary, Wisconsin (Pgs. 23 and 60)

Kelly Morgan: Baker River, New Hampshire (Pg. 81)

Mike Pagano: Humpback whales near Sitka, Alaska (Pg. 82)

Ron Vigil: Bear Lake and Salmon River, Idaho (Pgs. 20 and 106)

Roma Vigil: Metolius River, Oregon (Pg. 1); Palisades Reservoir, Wyoming (Pg. 89); and Table River, British Columbia (Pg. 143)

Bob Watts: Mousam River and Carrabassett River, Maine (Pgs. 29 and 59)

Ken Yates: Bosque Del Apache Wildlife Refuge, New Mexico (Pg. 24)

Internal graphic figures by Monica Klau

Introduction

Metolius River, Oregon

What is clean water? Clean water is a clear creek cascading down a steep mountainside, and a refreshing glass of ice water on a hot day. It is a spring-fed brook filled with wild trout, and rain falling on a parched field. It is a lush, green wetland teeming with vegetation and wildlife, and a dynamic estuary surging with the tide, filled with healthy shellfish and salmon. Clean water is all of these and more.

The quality of the earth's water is vital to our existence. We need ample clean water to quench our thirst, irrigate our fields, and sustain all life forms in the environment. We must have clean water in our homes, communities, businesses, industries, and in nature. We need clean water today and we will need it tomorrow.

We rely on clean water in almost every aspect of our lives. We rely on it for drinking, bathing, cooking, swimming, fishing, and boating. We count on it for growing and processing our food and nourishing the plants and animals. We count on the aesthetic qualities of clean water to nourish our souls.

Unfortunately, we have no guarantee that clean water, relied on so heavily, will always be available. The supply of clean water on the earth is finite, and it is being threatened by water pollution.

Water pollution is a serious problem today, in spite of our efforts to control it. The Environmental Protection Agency (EPA) estimates that approximately one third of all the waters in the

Gardiners Bay, Long Island, New York

United States are unsafe for swimming, fishing, and drinking. Many of these waters are suffering the effects of indirect or diffuse discharges of pollutants associated with stormwater runoff from adjacent lands. We call this type of water pollution *nonpoint source pollution* to differentiate it from *direct* or *point source discharges* of pollutants into waterways from pipes and outfalls. The polluted waters in the United States include our major waterways and their tributaries. The Mississippi River, for instance, which drains about half of the continental United States, has serious water pollution problems. Water quality degradation caused by erosion and sedimentation, municipal and industrial discharges, and agricultural runoff threaten its fish and wildlife. Structures constructed on the Mississippi for navigation and flood control also contribute to the decline in water quality.

The Columbia River, which drains most of the northwestern United States and parts of British Columbia, also has water pollution problems. Municipal and industrial discharges and agricultural runoff deliver a wide variety of pollutants to the river. Some of these pollutants, like dioxins and pesticides, are toxic in extremely small concentrations. Dams constructed on the Columbia River for hydropower and irrigation have altered water quality and fish habitat, contributing to the near extinction of some populations of salmon. The spring Chinook salmon run on the Columbia, which once numbered one hundred thousand fish or more, dwindled to an estimated ten to fifteen thousand in 1995, a decrease of almost ninety percent. The Snake River sockeye salmon, which migrate up the Columbia River, have fared even worse. In 1994, only one wild sockeye salmon returned to its spawning ground in Redfish Lake, Idaho.

Tributaries of these major waterways, such as Oregon's Tualatin River—a tributary of the Columbia—also are threatened by water pollution. The Tualatin River has high nutrient concentrations due to municipal and industrial discharges, stormwater runoff, and natural soil conditions. These high concentrations of nutrients, combined with warm water temperatures, cause unsightly algal blooms and unhealthy, fluctuating oxygen levels.

Bonneville Dam on the Columbia River, Oregon

Even greater problems exist in other parts of the world, where water quality has been a lower priority until recently. Parts of the Baltic Sea in Eastern Europe, for example, contain almost no fish or other aquatic life because of water pollution. Industries and municipalities in Germany, Poland, Denmark, and other surrounding countries have discharged untreated or partially treated wastes and wastewaters into the Baltic Sea for decades. Now, parts of the Baltic Sea's water and sediment are severely contaminated and few organisms can survive. In southern Poland, the Vistula River and others like it are highly contaminated with pollutants from mining and other heavy industrial activities. In other parts of Poland, rivers are contaminated with pollutants from food processing plants, textile mills, and municipal wastes. Many municipalities and industries still discharge untreated wastes and wastewaters into Poland's rivers and streams.

But there is hope. Until a few years ago, twenty million gallons of untreated municipal and industrial wastewater were pouring into the Rio Grande near Nuevo Laredo, Mexico every day. These untreated discharges were contaminating the water, sediment, and fish downstream of Nuevo Laredo with toxic compounds. In 1996, the community finally constructed a wastewater treatment plant and began treating these waste streams to remove pollutants prior to discharge.

Clearly, we cannot take clean water for granted. It is crucial to our survival, prosperity, and happiness and it is being threatened in all parts of the world. Moreover, the water environment knows no political boundaries. Water and the pollutants in it move freely across borders. We must address the world's water pollution problems not only as individual states and nations, but also as members of a greater worldwide environmental community.

Most people are concerned about clean water, yet may feel uninformed. They may not know about the many sources of water pollution or the methods used to prevent and control it. They may not be aware of the rules and regulations adopted to protect our water. They could be intimidated by the science of water pollution control. Many of us know how important clean water is, but we may not know how to get involved to help protect it.

This book is for you if you share some of these thoughts. Whether you have a technical or nontechnical background, you can use it as a reference to educate yourself broadly about water quality and water pollution control. It provides the answers to your questions in a clear and understandable way, covering a wide range of topics without the equations commonly found in textbooks. It contains straightforward explanations with additional references for those interested in exploring specific topics in more detail.

This book can help you learn about the water environment and how water moves through it continually. You can discover how the natural characteristics of rivers, lakes, wetlands, and oceans influence their quality. You can learn the basics of water chemistry and microbiology to help you understand the causes of water pollution and techniques used to prevent and control it. You can learn about water quality rules and regulations, drinking water, and ways of protecting water quality by looking at the small details and the big picture. Finally, you can discover how you can personally help protect water quality in your home and community to ensure that we have clean water now and in the future.

I. The Water Environment

Hood Canal, Washington

Where does the water environment begin? It begins where a single drop of rain falls to earth. This raindrop joins with others like it to form tiny trickles. These trickles combine and run off the land to create rivulets, creeks, streams, and rivers. The small streams and mighty rivers of the world unite to produce the vast oceans and seas that surround us.

The water environment is the entire world of water that we know and love: from cool, clear mountain streams to the dynamic and salty oceans. It is the loud cascading creeks and waterfalls and the quiet, slow moving rivers and peaceful lakes. It is the wetland waters filled with lush, green vegetation and the sterile melt waters running from glaciers across barren landscapes. The water environment also includes secret waters—springs, seeps, and groundwater—and the often-invisible atmospheric water poised to fall on the earth in a sudden cloudburst.

The Hydrologic Cycle

Although the world of water is enormous and all encompassing, it is connected in a single natural cycle. Water moves from the ocean to the land and back to the ocean again continuously. This cyclic movement of water through the environment is called the hydrologic cycle. It begins as water moves from the ocean's surface into the air above through evaporation. During evaporation, only the fresh water vapor and other volatile compounds enter the atmosphere. Minerals, salts, and other impurities are left behind in the ocean. The buildup of these minerals and salts over time has made the ocean salty.

Water evaporates into the atmosphere and forms clouds above the ocean. The prevailing ocean winds blow these clouds of moist air inland and as they rise to move over the mountaintops, the air in them cools. Because cold air cannot hold as much moisture as warm air, water falls from the clouds as rain or snow.

The moment a raindrop strikes the surface of the earth, it begins its journey back to the sea. Sometimes the raindrop soaks into the earth and moves slowly into the groundwater. Sometimes it runs off the land surface and moves quickly in a swift-flowing stream. Other times the raindrop rests in deep river pools or lakes, is taken up by plants and animals, or enters the atmosphere again through evaporation. Ultimately, the raindrop makes its way back to the ocean, which is like a giant reservoir. Water is stored in the ocean until it is delivered to the land as a result of evaporation and precipitation. Once the water reaches the land, it begins making its way back to the ocean through groundwater or surface water flow, and the cycle continues.

The hydrologic cycle

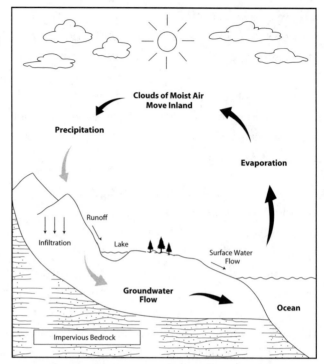

Clouds of Moist Air Move Inland

Precipitation

Evaporation

Runoff

Infiltration

Lake

Surface Water Flow

Groundwater Flow

Ocean

Impervious Bedrock

All the waters of the world are connected by the hydrologic cycle. The rivers and streams are connected to the lakes, ponds, and wetlands. The surface water is connected to the groundwater. The rivers are connected to the bays and estuaries, which are connected to the oceans and seas. These connections are extremely important to water quality; they allow materials entering the water at any point in the hydrologic cycle to move from one water body to the next. For instance, precipitation falling on exposed mine tailings high in the mountains may pick up and carry contaminants from the tailings into nearby streams, downstream lakes, and groundwater. These contaminants may show up many miles from their point of origin, possibly polluting a community's drinking water supply.

Although water takes on many different forms as it moves continuously through the hydrologic cycle, the world's supply of water is finite; we cannot make any more of it. What we have now is all we get. Because it is finite, protecting its quality is crucial. Our very survival depends on it.

Natural Conditions That Influence Water Quality

The different forms water takes in nature—the rivers and streams, lakes and ponds, wetlands, bays and estuaries, oceans and seas, and groundwater—all have unique water quality characteristics. These characteristics are influenced by the activities of humans and also by natural conditions in the environment. Some of the more important natural conditions include geology, climate, the amount and type of vegetation present, morphological characteristics such as the size, shape, depth, and width of water bodies, and the location of the water on the earth's landscape.

Geology

The geology of an area determines, in large part, the mineral makeup of its waters. For instance, water in areas with limestone deposits contains limestone minerals such as calcium and magnesium. These minerals dissolve and enter the water when it passes over rock formations and soil containing limestone. Water also will pick up small concentrations of metals such as copper, lead, and zinc when it passes over rocks and soils containing these elements. Because all minerals dissolve to some extent in water, you can discover much about the mineral content in water in any

Pacific Ocean near Garibaldi, Oregon

given area by learning what kind of minerals are found in the area's soil and rock.

Climate

Climate influences water quality because temperature, precipitation, and wind affect the physical, chemical, and biological characteristics of water.

Temperature is one of the most important natural conditions influencing water quality. It affects the amount of dissolved gases, such as oxygen, in the water. Warm water contains less oxygen than cold water, making it difficult for some organisms to survive. Also, chemical and biological reactions occur more rapidly in warm than in cold water, resulting in stress on some aquatic organisms. Chapter 2 provides more detail about the influences of temperature and dissolved oxygen on water quality.

The amount of precipitation falling in an area determines the number and size of its water bodies. Fewer water bodies exist in dry climates, and those that do tend to be smaller and more susceptible to pollution. Small bodies of water are more likely to become polluted than large bodies because they have less water available for diluting the effects of pollutants. In other words, an equal amount of pollutant would cause considerably more damage if discharged into a small creek than if discharged into a large river, simply because of dilution.

Wind is responsible for mixing the surface of waters, helping to enrich them with important gases like oxygen and carbon dioxide. Wind also influences the rate of evaporation from the surface of the water.

Vegetation

The presence or absence of vegetation also influences the natural quality of water. In areas where it is abundant, vegetation falls into the water, mixes with it, breaks apart, decomposes, and becomes part of the water. In some cases, excessive decaying vegetation can color the water. For example, one of the tributaries of the mighty Amazon River in South America is the color of ink because of decaying organic material. It is called the Rio Negro, or Black River, for obvious reasons. The water in wetlands is often a rich, brown, tea-like color because of decaying vegetation. In areas where vegetation is not abundant—in high mountain areas above timberline, for example—water contains less natural organic material. Sometimes waters located above timberline will be crystal clear and almost sterile, containing few minerals and nutrients and few fish or other aquatic organisms.

Vegetation such as trees and shrubs growing along stream corridors helps maintain desirable levels of dissolved gases in the water by shading it and keeping it cooler. Vegetation also acts as a filter to remove solid particles that are suspended in the water and helps to bind soil particles together to prevent erosion.

Morphology

The shape and dimensions of water bodies have a direct influence on their quality. For example, a shallow lake will generally be mixed thoroughly by the action of waves and wind. This mixing action helps to distribute minerals and dissolved gases equally throughout the lake. In contrast, a deep lake generally will not be well mixed. The bottom of the lake may have less oxygen and more minerals than the surface of the lake. Deep, unmixed lakes can develop layers, each with different water quality characteristics. (Stratification is the term used to describe this layering effect.)

Because streams on steep slopes flow swiftly, they often have better water quality than streams on gentler slopes. Streams on steep slopes experience more turbulence as water cascades over rocks and logs, adding oxygen to the water by mixing with the air. Streams located on mild slopes do not have the benefit of turbulent mixing to aerate the water. Swift-flowing streams, however, also have greater energy for causing erosion. Sediment from eroded stream banks may become suspended in the water, increasing turbidity and lowering the quality of the water.

Location

The location of a water body on the earth's landscape determines the natural conditions described above—geology, climate, vegetation, and morphology—and thus the natural quality of its water. As in real estate, location means everything. For instance, a slow-moving river meandering through a broad, flat valley will not have the same quality as a high mountain stream. The high mountain stream will likely be clear and cool while the valley river may be turbid and warm simply because of location and natural conditions.

We can see now that, even without humans, each of our water bodies would have different characteristics because of natural conditions in the environment. Unfortunately, the activities of humans tend to compound these natural differences, giving rise to many of the concerns we have about water quality.

Human Activities That Affect Water Quality

Many human activities threaten water quality. Some of these activities have been occurring for many generations and some began more recently. This section reviews these activities in relationship to the different water forms in the environment.

Lochsa River, Idaho

Rivers and Streams

Rivers and streams are the highways of the water world. People have used them to transport themselves and their goods from the mountaintops to the seas for centuries. Unfortunately, humans have used them also to dispose of and transport their wastes, a practice that seriously threatens water quality in our rivers and streams.

Since ancient times, villages have been built on riverbanks. Wastes from these villages were thrown into the rivers to be carried away. At first, few people lived downstream and the rivers had the natural capacity to assimilate the waste and cleanse themselves. This natural capacity for a water body to cleanse itself is called assimilative capacity. As the population continued to grow, however, the assimilative capacities of the waters were over-burdened and the rivers could no longer cleanse themselves.

Today, most of us know it is unacceptable to discharge untreated waste into a river or stream. Waste dumped into a river upstream will be carried downstream to the users below. The phrase "we all live downstream" is often used to remind us to use our rivers wisely, respecting the rights of all downstream users. In turn, we hope the people living upstream from us will respect our rights.

Although wastewater from most communities and industries is now routinely treated to remove pollutants, ultimately it is discharged into our rivers along with any pollutants that remain after treatment. Our efforts to keep rivers clean and healthy compete with this age-old practice of using our rivers to transport wastes.

Sometimes wastes enter our rivers and streams through more spread out, indirect, or diffuse discharges, or nonpoint source discharges. For instance, fertilizers, pesticides, and herbicides can be carried from our lawns and fields into nearby waters during and after rainstorms, as a result of stormwater runoff.

Lakes and Ponds

Our ancestors also established settlements on the banks of lakes and often discharged their wastes directly into them. Today, most of us realize that disposing of waste directly into a lake is a poor practice because it causes water pollution. Unfortunately, this practice continues in some areas and only recently has been discontinued in others. As an example, raw sewage was poured into Dal Lake high in the Himalayan Mountains during a recent civil war. The military turned resort hotels located along the shoreline into encampments and discharged untreated sewage directly into the lake, causing it to become highly polluted.

In Oregon, one of the more progressive environmental states in the United States, two municipalities were required to stop discharging their treated wastewater into downstream lakes only recently. Pollutants remaining in the wastewater, even after treatment, were harming the lakes. State environmental officials worked with community leaders to educate them about water pollution problems caused by these discharges. They also helped the communities find other means of disposal. Chapter 2 includes additional information about the problems associated with discharging nutrients into bodies of water.

Direct discharges into lakes and ponds also occur from stormwater runoff. Since stormwater picks up pollutants as it runs across the surface of the land, all activities occurring on the land surrounding a lake have the potential to contribute pollutants. For instance, many people remain interested in having a home or summer cabin at the edge of a lake. Unfortunately, residential development often results in both direct and indirect discharges of wastes and wastewater into our lakes. Direct discharges of fertilizers, pesticides, nutrients, and sediment may result from stormwater running off properties surrounding the lake. These direct discharges may also include materials from improper car and home maintenance such as gas, oil, antifreeze, soaps, and paints.

Indirect discharges may occur when homes are built around a lake where no community sewer system is in place. People often use septic systems in these areas. Unfortunately, sewage from the septic tanks and their drain fields may seep into the groundwater and move with the groundwater into the lake, causing contamination. This pattern of polluting lakes by improper use of septic systems has occurred across the United States and abroad.

Wetlands

Wetlands are truly rich. They are rich in nutrients, animal life, and vegetation. They support an abundant and diverse population of plants and animals, providing habitat for many species of aquatic vegetation and serving as a spawning ground and nursery for many species of fish. In fact, approximately one third of the plant and animal species listed as threatened and endangered in the United States depend on wetlands for habitat.

Wetlands connect the upland world with the world of open water, while providing a protective buffer or transition zone between the two. They protect the uplands from erosion by absorbing the effect of waves on the shoreline of open water. They also protect open water from upland disturbances.

Wetlands are also nature's filters. They filter out pollutants as water moves from upland into open water bodies. (Chapter 4 includes additional information about this filtration process and other natural methods of removing pollutants from water). Wetlands provide flood control and groundwater recharge zones,

Fresh water wetland, Tillamook County, Oregon

which are areas where surface water enters the groundwater and replenishes it.

If wetlands are so important, why is their existence in jeopardy? Until recently, society believed they were undesirable and unimportant. We have misunderstood them and considered them to be useless, unwanted swamps. Part of this misunderstanding comes from folklore, literature, and the popular media. Often, wetlands are portrayed as vile places were evil and mysterious events occur. For example, in *Oliver Twist*, Dickens associates evil deeds and unsavory characters with an area he describes as a low, unwholesome swamp bordering the river. Motion pictures like the Bogart and Hepburn classic *The African Queen* depict wetlands as undesirable places infested with mosquitoes and other pests.

These popular misconceptions have resulted in a threat to our wetlands even more serious than the threat to the quality of our rivers or lakes. Their very existence is in jeopardy. Wetlands continue to be lost throughout the world at an alarming rate. In the United States, for example, over ninety million acres of wetlands have been lost to date. Scientists estimate that only a little more than half of the wetlands that existed when European settlers moved to American are still in existence. In the past, many of our wetlands were lost because people drained them and turned them into agricultural properties. Today, the biggest threat is land development. As land values continue to increase and developable land and agricultural land becomes more scarce and expensive, the pressure to eliminate wetlands increases. We have only recently recognized how important wetlands are to the environment and enacted federal and state laws and local ordinances to protect them.

Bays and Estuaries

Bays and estuaries link the fresh water in our rivers and streams to the salt water in the ocean. These highly productive parts of the environment contain hundreds of species of plants and animals. They are dynamic, and fluctuate according to the movement of the tides and the changes in the fresh and salt water entering them.

On an incoming tide, saltwater enters the bay from the ocean and mixes with fresh water from upstream tributaries. The movement of saltwater and fresh water in and out of the bay in response to the tide and the inflows from tributaries creates brackish water—a mixture of salt and fresh water. Because of the constantly changing effects of the tide and tributaries, the characteristics of the water in a bay vary considerably with time and location.

The water quality in our bays and estuaries is threatened by upstream activities, since wastes discharged into our rivers and streams are carried into the bays and estuaries below. These waste materials may stay suspended while in the river because of the rapid movement and energy of the water, but when the river slows down as it reaches the estuary, the waste materials settle out.

Some of the activities that take place in our bays and estuaries also threaten their quality. For instance, the practice of storing and distributing petroleum products out of our bays and estuaries for transportation efficiency can result in contamination of both open water and shoreline if they are spilled or leaked. Also, boat maintenance activities such as stripping, sanding, and painting

Upper New York Bay, Ellis Island, New York

can harm water quality and aquatic organisms. For instance, shellfish growing in waters contaminated with tributilin (TBT), an additive used in boat paint to keep barnacles from growing on the hull, become deformed.

Many of the studies conducted on water pollution in bays and estuaries have focused on shellfish because of public health concerns and because of the economic importance of shellfish to coastal communities. Bacteria from animal and human wastes that enter the water have contaminated shellfish and have become a recurring problem in some areas. To protect the health of the organisms living in our bays and estuaries, as well as our own health, we must become aware of and begin controlling the activities that pollute them.

Oceans and Seas

Like bays and estuaries, our oceans and seas are forever changing. Their characteristics change due to climatic conditions and to movements of the moon and earth. Unfortunately, their characteristics have also changed for the worse because of human activities.

People have always discharged their wastes into the seas and they continue to do so. Because of their vast size, society has incorrectly assumed that oceans have an infinite capacity to assimilate waste materials. In recent years we have learned more about the finite nature of the oceans and the localized effects of pollution. For instance, medical wastes discharged into the Atlantic Ocean, including used needles and syringes, have washed up on beaches in the eastern United States. As previously mentioned, almost no fish or other aquatic life forms exist in parts of the Baltic Sea because industrial and municipal wastes have polluted the aquatic habitat.

Perhaps the biggest threat to the quality of water in our oceans and seas comes from oil spills. The 1989 Exxon Valdez disaster in the Gulf of Alaska (see Chapter 3) is one extreme example of this. The Gulf of Mexico has also been severely damaged by oil pollution. Water and sediment in the Gulf of Mexico and tributaries such as Mexico's Coatzacoalcos River are highly contaminated with petroleum products. This area, which supports the largest petroleum refinery complex in Latin America, has been the site of a number of disastrous oil spills. Petroleum spills in the Persian Gulf, the North Sea, and other parts of the world are also degrading water quality in our oceans and seas.

Indirect discharges also threaten water quality in the oceans. Because of the hydrologic cycle, all materials entering upstream waters that are not removed naturally or through treatment are discharged into the oceans. The oceans and seas are the ultimate sinks for all of the water on the planet and all of the pollutants dissolved or suspended in the water.

The oceans and seas are not infinite. They are particularly susceptible to the localized effects of pollution. We threaten their quality every time we discharge pollutants into them directly or indirectly.

Groundwater

Groundwater is water stored in the soil and rock formations below the earth's surface. It is the primary source of drinking water for many communities and the secondary source for others. Groundwater is used extensively for irrigation. It is also an important source of water for rivers and streams, especially during extended dry periods. Groundwater emerging at the bases of mountains and foothills provides the base flow of streams in the area during the dry season.

Groundwater provides the single largest supply of fresh water on the planet. It is used more extensively now than ever before because of society's increasing demand for fresh water. It is also being used more frequently today because drought conditions and contamination of surface water have reduced the availability of clean, fresh water at the surface.

As you learned earlier in this chapter, groundwater is connected to all other water forms in the environment through the hydrologic cycle. These connections make the threat of contamination to the surface water a threat to groundwater quality as well.

Bear Lake, Idaho

Activities taking place at the earth's surface are primarily responsible for groundwater pollution. For example, groundwater pollution can occur due to accidental spills and improper disposal of petroleum products and industrial solvents; over-application of fertilizers, pesticides, food wastes, and animal wastes to the land; and the use of septic systems in unsuitable locations.

Because groundwater is usually remote and inaccessible, it is difficult or impossible to clean once it becomes polluted. Methods of cleaning groundwater, such as isolating the contaminated area and pumping and treating the contaminated water, are not always successful. Regardless of success, attempts to clean groundwater are always expensive.

Atmospheric Water

As part of the hydrologic cycle, water evaporating from the ocean enters the atmosphere and then falls onto the land or back into the ocean as precipitation. This atmospheric water is the initial source of all fresh water in the environment.

Recall that when water evaporates off the ocean's surface, salts, minerals, and other impurities that do not evaporate are left behind. Thus, atmospheric water enters the hydrologic cycle in a relatively pure form. Unfortunately, the quality of the water in our atmosphere is threatened by air pollution. Air contaminated with sulfur and nitrogen compounds, for instance, mixes with water in the atmosphere and produces acids. When this contaminated atmospheric water falls as precipitation, it turns into acid rain, damaging and degrading the soil, vegetation, and water below.

Acid rain has been responsible for widespread environmental damage throughout the world, from the Appalachian and Adirondack Mountains of the eastern United States to the mountains of Eastern Europe and Scandinavia. Acid rain is particularly damaging to the aquatic environment. It has caused thousands of lakes and other water bodies throughout the world to become too acidic to support fish or other aquatic organisms.

Air polluted with carbon compounds contributes to another water quality concern: the greenhouse effect. The greenhouse effect, which is caused by atmospheric pollution insulating the earth and making it retain heat like a greenhouse, is thought to be warming the entire planet, including the polar ice caps and all the earth's waters. This condition is an important water quality concern because changes in the earth's temperature have a profound effect on all parts of the water environment. The temperature of the earth not only controls the melting and freezing of water, it controls the rates of chemical and biological reactions and the concentration of gases in the earth's water.

Summary

This chapter introduced you to the water environment and the natural cycle that connects all waters on earth: the hydrologic cycle. It introduced you to the natural conditions that influence water quality and to some important concerns about different waters. You can more fully appreciate these natural conditions and concerns if you understand some of the basic chemical and biological properties of water. The next chapter introduces you to many of the practical concepts in water chemistry and microbiology.

Additional Reading

Anikouchine, W. A., and R. W. Sternberg, 1981. *The World Ocean: An Introduction to Oceanography, Second Edition.* Prentice-Hall, Inc., Englewood Cliffs, New Jersey.

Burgis, M. J., and P. Morris, 1987. *The Natural History of Lakes.* Cambridge University Press, Cambridge, England.

Dahl, T.E., 1990. *Wetlands Losses in the United States 1780's to 1980's.* United States Department of the Interior, Fish and Wildlife Service, Washington, D.C.

Dunne, T., and L. B. Leopold, 1998. *Water in Environmental Planning (fifth printing).* W. H. Freeman and Company, USA.

Elsom, D., 1987. *Atmospheric Pollution.* Basil Blackwell, Inc., New York, New York.

Fitts, R.C., 2002. *Groundwater Science.* Academic Press, San Diego, California.

Leopold, L.B., 1997. *Water, Rivers and Creeks.* University Science Books, Sausalito, California.

Linsley, R. K., Kohler, M. A., and J. L. H. Paulhus, 1982. *Hydrology for Engineers, Third Edition.* McGraw-Hill, Inc., New York, New York.

Mitsch, W.J., and J.G. Gosselink, 2000. *Wetlands, Third Edition.* John Wiley & Sons, Inc., New York, New York.

Moran, J. M., Morgan, M. D., and J. H. Wiersma, 1986. *Introduction to Environmental Science, Second Edition.* W. H. Freeman and Company, New York, New York.

Pielou, E.C., 1998. *Fresh Water.* The University of Chicago Press, Chicago, Illinois.

Swanson, P., 2001. *Water the Drop of Life.* NorthWord Press, Minnetonka, Minnesota.

Todd, D. K., 1980. *Groundwater Hydrology, Second Edition.* John Wiley and Sons, Inc., New York, New York.

Van Dyk, J., 1995. *Amazon, South America's River Road.* National Geographic, Vol. 187, No. 2, February, 1995.

2. Water Chemistry and Microbiology

Lake Superior, Wisconsin

What is water made of? It is made of two simple elements: hydrogen and oxygen. When these elements are separate, they exist as colorless, odorless gases. When they are brought together, they form water vapor, liquid water, or ice, depending on temperature and pressure.

Water also consists of the materials dissolved or suspended in it, such as salts, minerals, and other dissolved substances, plus soil particles, debris, and other suspended solids. Water is alive. It contains small plants, animals, and microscopic organisms.

The basic concepts of water chemistry and microbiology introduced in this chapter will help you to better understand water quality and water pollution control. You do not have to be an engineer or a scientist, or be particularly good at math, to learn these concepts. This chapter presents them in simple and practical terms without using complex formulas and equations. However, additional references are listed at the end of the chapter for those interested in exploring specific topics in more detail.

Most of the terms used to describe the characteristics of water come from the disciplines of chemistry and microbiology. You may have heard some of these terms and wondered exactly what they meant. Reading this chapter will help you learn the language of water quality and water pollution control. You can also see the Glossary in the back of this book for definitions of useful terms.

Bosque Del Apache Wildlife Refuge, New Mexico

The Water Molecule

The smallest unit of water is called a molecule. It is made up of two atoms of hydrogen and one atom of oxygen, resulting in the familiar chemical term for water: H_2O. The chemical makeup of water gives it specific characteristics, such as its density and its ability to dissolve substances.

Density is a measure of the weight of a certain volume of a substance. For instance, a gallon of water weighs about eight pounds. The temperature of water helps determine its density. Cold water is denser, and therefore heavier, than warm water. This relationship is responsible for seasonal changes in water quality in some lakes. In the fall, the water on the surface of a lake cools and becomes denser than the water below, causing the surface water to sink. The warmer, lighter water on the bottom of the lake responds by rising to the surface, causing an overturn of the lake. This overturn results in a mixing of the suspended solids, nutrients, and dissolved gases in the lake's water (these substances are described later in this chapter).

Water is less dense when it exists as a solid than when it exists as a liquid. Thus, a gallon of ice weighs slightly less than a gallon of water. Water has this unusual property because water molecules expand when they freeze. Have you ever placed a water jug in the freezer only to have the water freeze, expand, and break the container? This property causes ice cubes to float in a glass of water and icebergs to float in the ocean.

The way hydrogen and oxygen are held together to form an individual water molecule, and the way each molecule is connected to the next, is called chemical bonding. The type of chemical bonding exhibited by water allows it to dissolve substances easily, making it a good solvent. Almost all solids, liquids, and gases placed in water will dissolve to some extent. Even solid copper and lead will dissolve slightly when placed in water or when water runs over rock or soil formations containing these elements.

Some of the substances that dissolve in water *reduce* its quality and some *improve* it. For example, when lead dissolves in water, it reduces its quality. When oxygen gas dissolves, it generally improves the water's quality and benefits the organisms living in it.

We often measure dissolved oxygen, temperature, and pH to characterize the quality of our water. These measurements tell us if the water is of sufficient quality for its intended uses. We also measure other substances to help characterize our water, including organic substances, inorganic substances, solids, nutrients, toxics, and microorganisms.

Dissolved Oxygen and Temperature

Dissolved oxygen is oxygen gas dissolved in water. Think of it in the same way you think of gas bubbles dissolved in a soft drink. The gas bubbles in a soft drink are bubbles of carbon dioxide injected into the solution during the carbonation process. Dissolved oxygen is oxygen gas that is entrained in the water as a result of water mixing with air containing oxygen.

Different types of water bodies contain different amounts of dissolved oxygen. A fast-flowing river usually contains more dissolved oxygen than a slow-moving one because it mixes rapidly with air containing oxygen while moving over rocks, logs, and debris in the stream. The highest concentration of oxygen is found in the whitewater stretches where the greatest amount of mixing occurs. Slow-moving rivers have less oxygen in them because they do not mix as rapidly with the air as they meander along.

Almost all water bodies contain dissolved oxygen, regardless of their turbulence. Even a still lake contains oxygen because oxygen from the atmosphere will dissolve on its surface. The transfer of oxygen from the atmosphere to the surface of a lake, or any other water body, is increased by the mixing action of wind and waves.

Dissolved oxygen is an important element in water because fish and most other aquatic organisms use it for respiration. Although fish don't breath the same way we do, their respiration process serves the same purpose. Fish take in water containing dissolved oxygen through their gills and circulate the oxygen through their bloodstream to provide energy. Just as humans cannot live without sufficient oxygen in the air, fish and other aquatic organisms cannot live without sufficient dissolved oxygen in the water.

Clackamas River, Oregon

Most cold water fish such as trout and salmon prefer dissolved oxygen to be in the range of about eight to twelve parts of oxygen per million parts of water, or more. If the amount of dissolved oxygen falls much below this range, the fish will, if possible, move to waters containing more oxygen, or else they will become stressed or perish. Oxygen is especially important to fish during their periods of reproduction.

Water quality professionals express the concentration of substances found in water in several different ways. Parts per million (ppm) can be understood by thinking of a container filled with a million marbles of equal size and weight. A reported concentration of ten parts per million of dissolved oxygen is equivalent to expressing that ten of the one million marbles are dissolved oxygen and the rest are water.

The other common way of expressing the concentration of substances in water is by stating the weight, or mass, of the substance found in one liter of water. For example, if the weight of solids in one liter of water is one hundred milligrams, the concentration is reported as 100 milligrams per liter (mg/L). Because one liter of water weighs approximately one million milligrams, the terms mg/L and parts per million are essentially

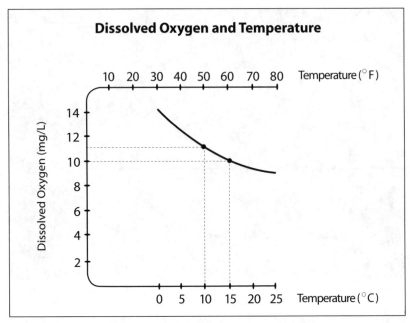

equivalent. In other words, reporting the concentration of dissolved oxygen as ten parts per million is the same as reporting the concentration as 10 mg/L.

As discussed in the last chapter, the amount of dissolved oxygen in a water body is affected by temperature. Oxygen is more soluble—meaning that it dissolves more readily—in cold than in warm water. One could say that cold water has a greater natural affinity for oxygen molecules than warm water. This relationship between temperature and dissolved oxygen causes a reduction in dissolved oxygen as the temperature of a fresh water body increases, as shown below.

For example, when a fresh water body is 10°C (50°F), the solubility of dissolved oxygen is 11.3 mg/L. If the temperature of the water body is increased to 15°C (59°F), the solubility is reduced to 10.1 mg/L. Most cold water fish prefer water with temperatures ranging from about 0°C to 15°C (32°C to 59°F). This temperature range generally provides them with a sufficient concentration of dissolved oxygen for their life processes.

When dissolved oxygen is at its maximum concentration, or solubility, at a given temperature, the water is saturated with oxygen. Most unpolluted, pristine waters—such as high mountain rivers and creeks—are ninety-five to one hundred percent saturated with dissolved oxygen. When waters become polluted, particularly with carbon and nitrogen compounds, they lose dissolved oxygen.

In addition to affecting the concentration of dissolved oxygen, water temperature also influences the rates of chemical and biological reactions, which generally increase with a rise in temperature. For example, bacteria and algae grow more rapidly in warm water than in cold water. This increase in the rate of reactions—which typically requires more oxygen, particularly for bacteria—coupled with a decrease in the amount of dissolved oxygen available, can stress aquatic organisms when water becomes too warm.

Because warm water can be harmful to many species of aquatic organisms, activities that increase the temperature of the water are generally undesirable. Today, industries that use large quantities

Mousam River, Maine

of water may employ cooling towers to cool the water heated in their production processes. This cooled water can be discharged into a nearby waterway without increasing the waterway's temperature. Foresters are beginning to modify their logging practices by leaving the trees that grow near streams to shade the water, keeping it cool. These uncut areas are called buffer zones because they buffer out, or minimize, the effects of pollutants. The term thermal pollution is used to describe discharges that cause undesirable shifts in water temperature.

pH

pH is an abbreviation representing the activity or concentration of hydrogen ions in a solution. It describes the acidic or basic (also called alkaline) condition of liquids on a scale that ranges from 0.0 to 14.0. Liquids having a pH of 7.0, such as distilled water, are neutral—neither acidic nor basic. Liquids with a pH lower than 7.0 are acidic. Strongly acidic substances are called acids. Liquids with a pH greater than 7.0 are basic. Strongly basic substances are called bases. Below are some common solutions and their approximate pH values.

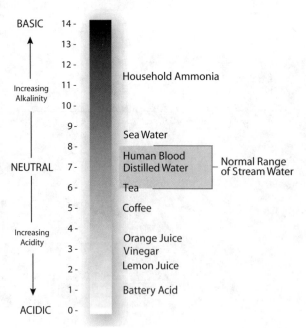

Acids and bases can be dangerous to humans and the water environment. They cause irritations and burning and can be extremely toxic to aquatic organisms. Fortunately, acids and bases have the ability to neutralize the effect of each other. For instance, you can neutralize an acid with a low pH by adding a base to it to bring it to a pH of about 7.0.

Many industries commonly use acids and bases in their production processes. The wastewaters generated from these processes are either acidic or basic and must be neutralized before being discharged in order to prevent water pollution. For instance, a company using acid in its production processes, such as a circuit board manufacturer, may generate an acidic wastewater with a pH of about 2.0. This industry would typically treat its wastewater by adding a base to neutralize the effect of the acid, obtaining a pH of approximately 7.0 before discharging it back into the environment.

Most rivers, lakes, and other natural water bodies have a pH ranging from about 6.0 to 8.5. The type and amount of dissolved minerals, gases, and aquatic organisms in the water determine the pH of water in nature. Most aquatic organisms cannot live if the pH of the water gets much outside of this natural range.

Organic Substances

Organic substances are materials made from carbon, such as plants and animals. All living organisms, from the largest tree to the smallest insect, are organic. All the foods we eat and all materials that are living or were once living, such as fallen timber, decaying vegetation, and petroleum products, are organic substances. Most organic carbon materials occur naturally; others are synthesized in the laboratory.

Many municipal, agricultural, and industrial wastes that are responsible for causing water pollution are organic. For instance, human wastes, animal wastes, and food processing wastes all consist primarily of organic materials.

Water quality professionals have developed special tests to measure the amount of organic material in water samples. The two tests they use most often are those that measure biochemical oxygen demand (BOD) and chemical oxygen demand (COD).

The BOD test is a laboratory procedure used to measure the natural process known as biodegradation. In biodegradation, small organisms that exist naturally in the environment—particularly bacteria—decompose organic substances by using them as a food source. In essence, the small organisms eat the organic substances. For example, if you put a banana peel out in your garden and leave it for a long time, it will eventually disappear into the soil. The living organisms, called *biota*, in your garden soil will *degrade* the organic banana peel, hence the term biodegradation.

To conduct the BOD test, a technician places a water sample in a small bottle about the size of a canning jar. The amount of dissolved oxygen in the water sample is measured and the bottle is placed in a warm room for five days. During these five days, naturally occurring bacteria—or "seed" bacteria added to the bottle—begin to biodegrade the organic material in the sample. As the bacteria decompose the organic substances, they use dissolved oxygen in the bottle for performing their life processes, such as respiration, metabolism, and reproduction. The technician measures the amount of dissolved oxygen in the water sample again at the end of the five-day period. The difference between the amount of oxygen at the beginning and end of the BOD test is

Laboratory BOD bottles

called the oxygen demand or biochemical oxygen demand. The test measures the amount of oxygen used to biodegrade the organic substances in the bottle, which is an indirect measure of the amount of organic material in the original water sample.

Because the test is conducted over a five-day period, it is often referred to as a five-day biochemical oxygen demand (BOD_5) test. Water quality professionals commonly use a five-day test period, in part, because of the history of water pollution control. When scientists and engineers began studying water pollution, they estimated it would take five days for water to travel from the mountains to the sea. They reasoned that this time period would be satisfactory for conducting their oxygen demand tests. This practice of conducting the test over a five-day period continues today.

The COD test also measures the amount of organic material in a water sample by measuring the oxygen demand. In the COD test, however, chemicals added to the sample—instead of bacteria— are responsible for breaking down, or oxidizing, the organic material.

Generally, the BOD test is used for municipal wastewaters and the COD test is used for industrial wastewaters. Organic substances measured as BOD or COD are important because they can affect the amount of dissolved oxygen in a stream in the same way they affect dissolved oxygen in the sample bottle. When organic substances enter a stream, they biodegrade. The bacteria in the stream use the organic materials for food and the dissolved oxygen in the water for energy and respiration. This process removes the dissolved oxygen needed by fish and other aquatic organisms from the stream. As mentioned earlier, the problem of having low levels of dissolved oxygen is compounded when the water is warm, because bacteria grow more rapidly and thus require more oxygen, but less oxygen is available.

When not managed properly, organic waste materials can kill fish and other aquatic species. For example, if animal waste from a cattle feedlot is not managed properly and is allowed to enter a waterway, the cattle waste will begin decomposing and strip oxygen from the water. Without oxygen, fish and many other aquatic organisms will not survive.

Inorganic Substances

Inorganic substances include rocks and minerals; metals such as gold, silver, copper, lead, zinc, and chrome; and solids like sand, silt, and clay. Many of the forms of nitrogen and phosphorus used for fertilizer are also inorganic substances.

Calcium and magnesium, two of the many inorganic substances found in rocks and minerals, are important because they are responsible for creating hard water deposits on plumbing fixtures and in industrial boilers. Waters with high concentrations of calcium and magnesium are called hard waters. Other inorganic substances found in rocks and minerals, such as sulfur compounds, are important because they can cause taste and odor problems in drinking water.

Humans use metals for many different purposes. We use gold, silver, and copper in jewelry. We use silver for photography. We use chrome for making car bumpers and tanning leather. We have used lead in paints and fuels. We commonly use metals in products ranging from automobiles to computers to kitchen appliances. Because we use metals so extensively, they can enter the water by many different pathways. Metals will dissolve slightly when placed in water, or when water runs over rocks and minerals containing metals. These dissolved metals can be toxic to aquatic organisms if their concentrations get too high.

Even in low concentrations, long-term exposure to metals can affect the health of aquatic organisms and humans. For example, human infants exposed to lead have abnormal brain development. Until recently, we used leaded gasoline, which caused relatively high concentrations of lead to build up on our highways. During rainstorms, this lead was washed off the roadway by stormwater runoff and entered nearby waters.

Because many metals and other inorganic materials are not very soluble in water, they are usually found on the banks or bottoms of water bodies or in solids suspended in the water.

Solids

Solids found in water consist of both inorganic and organic materials, such as soil particles and small pieces of vegetation, respectively. They are an important water quality concern because they cause turbidity, which can be harmful to fish because it reduces visibility. Also, the solids causing turbidity can be abrasive to their gills. Solids falling to the stream bottom can cover and harm bottom dwelling, or benthic, organisms. Excessive solids also make it difficult and expensive to treat drinking water and to disinfect wastewater.

Tests for total suspended solids (TSS) and total dissolved solids (TDS) help determine the amount and types of solids found in a water sample. A technician measures TSS by passing a sample of water through a clean filter, much like a coffee filter. The solids left on the filter after the water passes through it are the suspended solids. The technician weighs the filter before and after passing the water to determine the weight of the total suspended solids in the sample. The filter is dried before the final weighing so that the moisture in the filter is not included in the weight of the solids.

The technician collects the water passing through the filter in a glass container. Some of the solids in the water sample are so

Laboratory equipment for solids measurements

small they pass completely through the filter, but are captured in the glass container. The technician places the glass container on a burner and boils off the water. The solids left on the bottom of the container after the water boils off are so small they are difficult to see. However, wiping the bottom of the container with a clean paper towel picks up a slight residue. This residue consists of the dissolved solids in the sample. The technician weighs the glass container before and after the test to determine the weight of the TDS.

By adding together the TSS and the TDS, the technician obtains all of the solids in the sample, referred to as the total solids (TS).

The solids found in a water sample are reported in units of concentration. For example, if one liter of water is passed through the filter, and the suspended solids retained on the filter weigh four hundred milligrams, then the concentration of TSS is reported as 400 mg/L. If the dissolved solids passing through the filter weigh one hundred milligrams, then the concentration of TDS in the sample is reported as 100 mg/L. The TS concentration in the sample is the combination of these two values, or 500 mg/L.

When you read about the concentration of materials found in water, it may be difficult to get a sense of the quantities involved. To aid your understanding, try thinking about the small packets of salt and pepper found at some restaurants. Each of these packets contains about 500 milligrams of salt or pepper. If you emptied one of these packets of salt into one liter of distilled water, the concentration of salt in the water would be 500 mg/L. Now, suppose you add one 500-milligram packet of pepper to the same container filled with one liter of distilled water. You would then have 500 mg of salt and 500 mg of pepper in one liter of water. The salt would eventually dissolve and become dissolved solids. The pepper would not dissolve, so it would become suspended solids. A solids analysis of your one liter of seasoned water would result in a measurement of 500 mg/L of TDS (the salt), 500 mg/L of TSS (the pepper) and 1000 mg/L of TS (the salt plus the pepper).

Chapter 5 presents ways that total suspended solids and total dissolved solids are regulated by permits and water quality standards to protect water quality.

Nutrients

Nutrients are the elements all living organisms need for growth. They are the nutritional building blocks of bacteria, fish, trees, and humans. They are what we are made of. When farmers apply fertilizer to their fields, they are adding nutrients to help their crops grow. Nutrients come in both organic and inorganic forms. The most important nutrients in water quality and water pollution control are carbon, nitrogen, and phosphorus.

Carbon is one of the most abundant elements on earth. It can be found in plants, animals, soil, and the air we breathe. It is one of the key building blocks of all living things. Because it is so important, it is used by nature over and over again. It is recycled.

Bacteria are the best recyclers in nature. They remove carbon from organic materials through the biodegradation process, then use some of the recycled carbon to form new cells and release the rest into the atmosphere in the form of carbon dioxide gas. Green plants use the carbon in carbon dioxide gas as building blocks for their growth. This carbon recycling process occurs naturally in the environment and is essential for life on earth.

However, when too much carbon is released into our waterways in an uncontrolled manner, such as discharging untreated municipal sewage or industrial waste, it causes water pollution. Carbon in the discharged sewage or industrial waste is biodegraded and recycled in the water by naturally occurring bacteria. During biodegradation, bacteria remove dissolved oxygen from the water for energy and respiration. Because bacteria use up oxygen during biodegradation, they leave little for fish or other aquatic organisms. Recall that dissolved oxygen from the water is needed for respiration by aquatic organisms in the same way we need oxygen from the air. In summary, the uncontrolled discharge of carbon into the environment results in the removal of oxygen from the water, threatening the health of fish and other aquatic species.

One of the ways we prevent substances containing carbon from polluting our waters is by constructing municipal and industrial wastewater treatment plants. These treatment plants remove most of the materials containing carbon from the wastewater before it is discharged. Chapter 4 presents more detailed information about municipal and industrial treatment processes.

Nitrogen is also an abundant element in the environment and it is an essential nutrient for plant and animal growth. Bacteria recycle nitrogen through biodegradation in much the same way they recycle carbon. They use part of the nitrogen for cell growth and turn the rest into less complex forms of nitrogen, such as ammonia, nitrogen gas, and other inorganic forms of nitrogen.

One form of inorganic nitrogen that is responsible for problems in drinking water is called nitrate. Infants who drink water with too much nitrate in it—more than 10 mg/L—may be stricken with an ailment called methemoglobinemia, or "blue baby" disease. When nitrate gets into an infant's blood stream, it reduces the amount of oxygen carried by the red blood cells. Because it does not get enough oxygen, the infant turns blue and can die. Adults do not have the same problems with nitrate in their drinking water because their respiratory systems are more fully developed.

Like carbon, the uncontrolled discharge of nitrogen in the form of ammonia or organic nitrogen also can cause water pollution due to loss of oxygen from the water during the biodegradation process.

The other major nutrient of concern in water quality protection is phosphorus. Unlike carbon and nitrogen, phosphorus is not always an abundant element in the environment, yet it is necessary for the growth of all living things. When phosphorus is not available, organisms will not grow. For example, one of the reasons high mountain waters are so clear is that they lack enough phosphorus to support the growth of many aquatic organisms. When the absence of phosphorus or any other element limits biological growth, it is called the limiting nutrient.

Having too many organisms growing in our waters is generally undesirable. An abundance of aquatic organisms such as algae growing in a water body may cause it to turn green and turbid. In addition to being unsightly, algae can cause undesirable shifts in pH and dissolved oxygen in the water.

Lily pad pond, Columbia River Gorge, Washington

In waters where phosphorus is the limiting nutrient, the abundant and undesirable growth of algae and other aquatic organisms can be prevented and controlled by minimizing the amount of phosphorus available. To achieve this goal, some communities have banned or restricted the use of detergents made of phosphorus to reduce the amount of phosphorus discharged into their waters. Others have added special treatment processes to their sewage treatment plants to remove phosphorus. Farmers have been encouraged to limit their use of fertilizers to prevent excess amounts from being washed off of their property and into area streams.

When a water body is rich in nutrients, causing certain organisms such as algae to grow abundantly, it is eutrophic. When a water body contains few nutrients or has very low concentrations of nutrients and little biological growth, it is oligotrophic. When a water body is between these two extreme trophic conditions, it is called mesotrophic.

These trophic conditions can be natural states of water bodies related to their location on the earth's landscape (see Chapter 1). Water bodies that are away from natural nutrient sources tend to be oligotrophic and water bodies that are near abundant natural nutrient sources, or are subject to pollution, may be eutrophic. When the activities of humans add pollutants and nutrients to the water, we are contributing to eutrophication.

Toxics

The chemical substances most detrimental to water quality are the toxic compounds. In contrast to nutrients that increase biological activity, toxic substances can cause death and deformation of the organisms in the water.

One of the most commonly used toxic compounds is chlorine. Chlorine is used throughout the world as a disinfectant to kill harmful organisms and to protect human health. Unfortunately, the same characteristics that make it a good disinfectant—its ability to kill quickly in low concentrations—also make it harmful to the environment.

Chlorine is used to kill harmful organisms like bacteria and viruses at sewage treatment plants. However, it may also kill or harm desirable organisms such as fish and reptiles when the treated wastewater, called effluent, is discharged back into the environment. Chlorine toxicity can be avoided if the concentration of chlorine is kept low and if the effluent is discharged into a large, well-mixed water body.

Sometimes it is necessary to chemically remove chlorine from the effluent before discharging, to prevent toxicity. This dechlorination process is often accomplished by injecting sulfur dioxide gas into the effluent. The sulfur dioxide gas reacts with the chlorine molecules, converting them into nontoxic chloride molecules. Using other nontoxic methods to disinfect the wastewater, such as passing the effluent through ultraviolet light or adding ozone, can also prevent chlorine toxicity.

Chlorine is also used in the process of bleaching pulp to make paper white. During the bleaching process, chlorine molecules combine with organic molecules in the pulp to form harmful compounds called chlorinated organics. The most harmful of these chlorinated organic compounds are the dioxins. Members of the dioxin family of compounds have been shown to cause cancer in laboratory animals in extremely small concentrations. For instance, one type of dioxin called 2,3,7,8 TCDD has an established drinking water standard of 0.00000003 mg/L due to its harmful effects in low concentrations. (Chapter 7 provides more information about drinking water standards). Because of the toxicity of chlorinated organic compounds like dioxins, pulp and paper mills are working to reduce or eliminate the use of chlorine in their bleaching processes.

Ammonia is another common toxic compound. It is formed from the breakdown of organic materials containing nitrogen. It is found throughout the environment and in municipal and industrial wastewaters. Ammonia is not a concern in low concentrations (less than 1.0 mg/L), but it becomes toxic in high concentrations, especially in waters that are warm or have a high pH. Waters with these conditions cause ammonia to be in its most toxic chemical form: un-ionized ammonia.

Many other organic substances are also toxic. For example, petroleum products such as gasoline, diesel, fuel oil, and kerosene are all toxic organic compounds. Other toxic organic chemicals include widely used pesticides and herbicides such as alachlor, aldicarb, atrazine, chlordane, and 2,4-D, and organic solvents and industrial cleaners such as benzene and TCE. Chapter 5 describes how several of these toxic organic compounds are regulated to control the use and disposal of hazardous materials.

Toxic compounds can cause both short- and long-term health effects. If the health effects occur over a short period of time, the toxic compounds cause *acute* toxicity. If the health effects occur over a long period of time, the compounds cause *chronic* toxicity.

Laboratory tests called bioassays can be used to evaluate the toxicity of municipal and industrial effluents. In bioassays, technicians place small organisms such as minnows, algae, or water fleas in different concentrations of effluent. They then monitor the health of the organisms to determine what dose or concentration, if any, causes toxicity. Toxicity can be measured as weight loss, reduced reproduction, or mortality.

Microorganisms

The smallest living organisms on the earth are called microscopic organisms or microorganisms. These include bacteria, algae, and viruses. Although many microrganisms are too small to be seen with the naked eye, they are abundant in most natural waters. One gallon of river water may contain more than one million bacteria and more than ten thousand algae.

Bacteria are simple, single-celled organisms. They are so small you cannot see them with the human eye alone. With the advent of the microscope, however, scientists found that although they are tiny, they are vital to life on earth.

Long Island Sound, Nassau County. New York

Bacteria are the worker bees and recyclers of the water world. They are constantly breaking down complex compounds into simpler ones and recycling them back into the environment. For example, they will break down a complex sugar molecule into carbon, oxygen, and water. Bacteria use some of the components for energy and growth; they release the rest for use by other microorganisms. The recycling activities of bacteria are fundamental to water quality and water pollution control.

Many of the bacteria that affect water quality use carbon from organic molecules as their food source and as the building blocks for new cells. Many of them also use oxygen for respiration in much the same way humans do. Although bacteria are essential to recycling nutrients in the environment, and many of them are beneficial, some types of bacteria are responsible for human diseases. For example, certain species of bacteria are responsible for cholera, typhoid, dysentery, and other diseases that are transmitted in water.

Algae are ten to one hundred times larger than bacteria. They are made of single cells or multiple cells. The single-celled algae are generally too small to be seen without a microscope, but you can see the multiple-celled algae with the human eye. The simple blue-green algae are examples of single-celled algae. A seaweed is an example of multiple-celled algae.

Algae differ from bacteria in their energy requirements and growth mechanisms. Most algae are photosynthetic organisms, meaning they use light energy, such as energy from the sun, for growth. Because algae require light for growth, they are generally found only near the surface of most waters. Algae use carbon dioxide from the atmosphere and the water as carbon building blocks for creating new cells.

Algae grow rapidly in polluted waters that contain an abundance of nutrients. These waters may have an unsightly green sheen to them, which is characteristic of the presence of algae. Algae can also cause undesirable shifts in water chemistry, such as fluctuations in dissolved oxygen and pH in the water at different times of the day, depending on the amount of available light. These undesirable fluctuations may stress or kill other organisms living in the water.

Viruses are the smallest microorganisms. They are ten to one hundred times smaller than bacteria. They are called parasites because they cannot live outside the cell of another organism. The organisms they live in are called the host organisms. Bacteria are often the host organisms for viruses. Viruses are responsible for causing human diseases such as smallpox, infectious hepatitis, influenza, yellow fever, and poliomyelitis. Viruses have also been linked to some types of cancer. Of these diseases, scientists currently believe that only infectious hepatitis is caused by the transmission of viruses through the water.

Summary

This chapter introduced you to some of the basic principles of water chemistry and microbiology that are fundamental to water quality and water pollution control. You learned about the water molecule, dissolved oxygen and temperature, pH, organic and inorganic substances, solids, nutrients, toxics, and microorganisms. Now that you have familiarized yourself with these concepts, you can begin to more fully understand the sources of water pollution and the techniques used to control it, as described in the next two chapters.

Additional Reading

American Public Health Service, et al., 1998. *Standard Methods for the Examination of Water and Wastewater, 20th Edition*. Published jointly with the American Water Works Association and Water Environment Federation, Washington, D. C.

Baird, R.B., and R.K. Smith, 2002. *Third Century of Biochemical Oxygen Demand*. Water Environment Federation, Alexandria, Virginia.

Benjamin, M., 2002. *Water Chemistry*. McGraw-Hill, New York, New York.

Bitton, G., 1994. *Wastewater Microbiology*. John Wiley & Sons, Inc., New York, New York.

Gaudy, A. F., and E. T. Gaudy, 1980. *Microbiology for Environmental Scientists and Engineers*. McGraw-Hill, Inc., New York, New York.

Hem, J. D., 1992. *Study and Interpretation of the Chemical Characteristics of Natural Water, Third Edition*. United States Geologic Survey Water-Supply Paper 2254.

Hemond, H.F., and E.J. Fecher-Levy, 2000. *Chemical Fate and Transport in the Environment*. Academic Press, San Diego, California.

Kegley, S.E., and J. Andrews, 1998. *The Chemistry of Water*. University Science Books, Sausalito, California.

Laws, E. A., 1993. *Aquatic Pollution: An Introductory Text*. John Wiley & Sons, Inc., New York, New York.

Maier, R.M., and I.L. Pepper, 2000. *Environmental Microbiology*. Academic Press, San Diego, California.

Manahan, S. E., 2001. *Fundamentals of Environmental Chemistry, Second Edition*. Lewis Publishers, CRC Press, Inc., Boca Raton, Florida.

Masters, G. M., 1991. *Introduction to Environmental Engineering and Science*. Prentice-Hall, Inc., Englewood Cliffs, New Jersey.

Sawyer, C. N. and P. L. McCartey, 1978. *Chemistry for Environmental Engineering, Third Edition*. McGraw-Hill Book Company, New York, New York.

3. Sources of Water Pollution

Weber River, Utah

What causes water pollution? This question has many answers. Humans and the various activities we participate in cause water pollution, as do forest fires, floods and other natural events. A catastrophe such as the wreck of an oil tanker can pollute our waters, but so can a less dramatic event like a failing septic system.

Some of the most common causes of water pollution include stormwater runoff, domestic discharges, industrial discharges, accidental spills, and use of water control structures such as dams.

Stormwater Runoff

When rain falls on roads, parking lots, or farm fields, it either soaks into these surfaces or runs off. The rainwater that runs off is called stormwater runoff.

Stormwater runoff typically contains many pollutants. For example, runoff from roadways and parking lots often contains oil, gasoline, and other automobile fluids. Runoff from farm fields may include pesticides, fertilizers, and animal wastes. Runoff from forested areas may have soil, vegetation, and other debris in it. Stormwater runoff from golf courses may contain pesticides and fertilizers. Industrial sites may produce runoff containing industrial chemicals. Runoff from construction or other areas where the land is being disturbed often contains eroded soils, dissolved minerals, and debris.

Stormwater runoff carrying these pollutants enters storm gutters, pipes, and ditches—and ultimately our rivers, streams, and lakes. We only recently began treating stormwater runoff to remove pollutants before it enters our waterways. (Chapter 4 describes methods of treating stormwater.)

These sources of stormwater pollution, alone and combined, have resulted in serious water pollution problems. Stormwater discharges from agricultural fields, for instance, have contaminated rivers with organic material and bacteria from animal waste. Stormwater discharges containing phosphorus have contributed to the undesirable growth of large numbers of algae in lakes and rivers. Stormwater discharges containing eroded soils have damaged fish spawning beds and other aquatic habitat.

Most of the sources causing stormwater-related pollution are referred to as nonpoint sources because they come from broad areas rather than single points of origin. How big is the problem of nonpoint source pollution? According to the Environmental Protection Agency (EPA), nonpoint sources of pollution are the

Manhattan Bridge and East River, New York

leading cause of water pollution in the United States today. Only about half of the miles of rivers and acres of lakes assessed by state agencies have sufficient water quality to fully support their designated uses for fishing, swimming, and drinking. Nonpoint pollution from agriculture, in particular, has been cited by the EPA as a major cause of the degradation of these waters.

Other sources of pollution are referred to as point sources because they originate from single points of discharge, such as the ends of pipes. Domestic discharges are point sources of water pollution.

Domestic Discharges

Most of us take modern sewage collection and treatment for granted. We flush the toilet and the waste goes away. Where does it go? In most cases, it goes either to a septic system or to a sewage treatment plant.

Septic Systems

Septic systems consist of a large buried tank, usually about 1,000 gallons (3,800 liters) in size, and a series of perforated pipes—called a drain field—that are placed in the soil down slope from the tank. Sewage solids are retained in the tank and the liquid effluent enters the drain field and percolates into the underlying soil. Chapter 4 outlines the proper use of septic systems in more detail.

Millions of people in the world rely on septic systems for managing their sewage, making these systems a considerable source of water pollution when used inappropriately. Septic systems can cause water pollution when they are placed in areas with poor soil conditions, high water tables, or in areas without sufficient area for them to function properly. For example, septic systems do not work well when placed in tightly packed, fine-grained soils such as clay, because effluent from the septic tank cannot pass through the soil easily. Instead, it collects at or near the surface of the ground and may run off into nearby waters.

Effluent from a septic tank is essentially raw sewage and it poses both an environmental and a health hazard. If drain pipes from the septic tank are too close to the water table, effluent will enter the groundwater before being properly treated in the soil, resulting in groundwater contamination. If septic systems are placed too close together in densely populated areas, adequate space will not be available for the systems to function properly. The soil will become overloaded beyond its capacity to adequately treat the wastewater, resulting in surface or ground water pollution.

Sewage Treatment Plants

Fortunately, in the United States, municipal sewage collection and treatment plants serve most areas where housing is dense. In areas where cities provide sewer service to their residents, wastewater from individual homes flows through house plumbing and underground sewer pipes to the community's sewage treatment plant. The plant treats the wastewater to remove approximately eighty-five percent of the solid and organic materials in the wastewater, disinfects it to kill bacteria and viruses, and then usually discharges it into the nearest waterway. Chapter 4 describes the operation of sewage treatment plants in more detail.

Although sewage treatment plants are intended to help prevent water pollution, they can also contribute to it in a number of ways. The solid and organic materials not removed through the conventional treatment process, which is approximately fifteen percent of the amount entering the system, can degrade water quality. For instance, when large volumes of effluent are discharged into small, poorly mixed waterways, wastewater can dominate the receiving stream, reducing the concentration of dissolved oxygen in the stream.

Nutrients not removed through the conventional treatment process, especially phosphorus, can contribute to eutrophication, causing the undesirable growth of algae and other nuisance organisms. Nitrogen, in the form of ammonia, can be toxic to fish and other aquatic species and can also lower dissolved oxygen in the receiving stream.

Chlorine, which is commonly used for disinfection at a treatment plant, can be toxic to organisms in the receiving water. Again, the toxic effect of chlorine is more pronounced in small waterways with low flow rates and poor mixing conditions.

Conventional municipal treatment systems do not typically treat toxic substances very well. Substances such as some household cleaners, petroleum products, metals, and other toxic compounds that are sometimes found in sewage can pass through the system without being properly treated, causing toxicity in the receiving stream. They can also contaminate the sewage solids removed during treatment, making these solids difficult to dispose of or use for beneficial purposes such as fertilizer without harming the environment.

Poor performance of the treatment system due to upsets in the biological treatment process (see Chapter 4), operator error, or storm-related problems can result in the discharge of improperly treated sewage. Poor performance under stormy conditions is a serious problem for both municipal treatment plants and collection systems. During a rainstorm, stormwater can enter sanitary sewer pipes through cracks or improper joints. This stormwater flows to the treatment plant, where it can overload the system. These extra flows reduce the time available for settling and biological treatment, and can result in insufficient treatment of the sewage prior to discharge.

In some older communities, the pipes carrying sewage are connected to the pipes carrying stormwater. During rainstorms, the sewer pipes become so full they overflow into the stormwater pipes. These stormwater pipes do not go to the sewage treatment plant; they go directly to nearby waterways. These overflow events are called combined sewer overflows (CSOs). Combined sewer overflows result in untreated sewage being discharged directly into a nearby river or stream, posing both an environmental and a human health hazard.

CSOs are a serious problem in larger, older cities where these combined sewage and stormwater collection systems were used extensively in the past to save money. Cities in the United States currently contending with CSO problems include Portland, Oregon; Seattle, Washington; New York, New York; Boston,

Combined sewer outfall, Willamette River, Oregon

Massachusetts; Chicago, Illinois; and St. Louis, Missouri. These cities and others are currently spending millions of dollars to separate their sewage and stormwater collection systems. They are also trying to discover other ways of solving their CSO problems, such as treating the combined wastewater prior to discharge.

Industrial Discharges

Although domestic discharges can be a significant source of water pollution, they usually pose less of an overall threat to water quality than do industrial discharges. Industrial discharges are often larger, and they may contain more harmful materials.

Large, "wet" industries like pulp and paper mills are almost always built near the banks of rivers because of their high demand for water. They obtain clean water by directing river water from upstream diversions to their plants below. Once the clean water reaches the plant, it is used in the production process and then sent to a treatment system to remove the pollutants acquired during production. This treated process water is then discharged back into the river downstream of the plant. Obviously, the process

East River near Brooklyn Bridge, New York

waters used by industry must be treated properly to prevent water pollution. For instance, at a pulp and paper mill, clean water is mixed with wood chips and chemicals during the pulping process. This processing water must then be treated to remove the pollutants acquired from both the wood and pulping chemicals prior to discharge.

Industrial treatment systems generally remove more than ninety percent of the solid and organic materials in the wastewater. They also address other industry-specific problems by removing metals or neutralizing acids. However, like municipal systems, industrial treatment systems can also contribute to water pollution, though they are intended to prevent it. Water pollution problems can arise from: 1) improper operation and poor performance of the treatment system, 2) the adverse conditions caused by the ten percent or more of solids, organics, and other materials that cannot be not removed through standard treatment processes, and 3) specific chemicals that are toxic in low concentrations and difficult to remove entirely.

Which industries use water in their production processes? Almost all industries manufacturing any type of product use water during production. Some of the more common industries requiring process waters include food processors, electronic equipment manufacturers, rare metal manufacturers, forest products producers, textile manufacturers, pharmaceutical manufacturers, pulp and paper mills, leather tanners, and chemical manufacturers. These industries and the many others that use water in their production processes must continuously provide proper treatment of their process waters prior to discharge to prevent water pollution.

Accidental Spills

Industrial and domestic discharges and stormwater runoff are somewhat predictable sources of water pollution. When it rains, we know stormwater pollutants run off the land and enter our waterways. When domestic and industrial discharges are not managed properly, we know they can degrade the quality of our waters. Another source of water pollution results from events that are unpredictable: accidental spills.

In the early hours of March 24, 1989, the thousand-foot long oil supertanker Exxon Valdez traveled outside of its shipping channel and ran aground on Bligh Reef in Prince William Sound, Alaska. This catastrophe resulted in the largest and most devastating oil spill in United States history. Almost eleven million gallons of crude oil spilled into the sound. The spill migrated over an area of more than 11,000 square miles, killing an estimated 300,000 birds and hundreds of mammals. The spill killed bald eagles, peregrine falcons, murres, loons, grebes, and puffins. It also killed and injured sea otters, harbor seals, salmon, and many other species.

The effects of this tragic spill could have been prevented or lessened in many different ways. Those responsible for piloting the ship could have been more conscientious and more skillful in keeping the ship on course. The tanker could have been double-hulled to provide a secondary means of containment once the first hull ruptured. The emergency response could have been more rapid and better organized. Official crews did not arrive on the scene for approximately ten hours and it took them more than thirty-seven hours to place the floating booms used to corral the floating oil. (Chapter 4 describes spill prevention and cleanup techniques used to prevent and control water pollution.)

Although this major oil spill was devastating to the environment, nature is now recovering in Prince William Sound. Much of the oil floating on the surface of the sound migrated into small coves and inlets, protected from wind and waves, where it eventually congealed and fell to the bottom of the sound. Oil that had washed up on the shoreline adhered to soil and vegetation, where it formed sticky tar and asphalt-like deposits. Given enough time, these deposits of spilled oil will eventually be biodegraded by bacteria and other naturally occurring microorganisms. Although we know that biodegradation will eventually occur, no one knows how long it will take. Neither the long-term effects of the spill nor the environment's ability to recover will be known for many years.

Many different types of materials have spilled into waterways in the United States and continue to do so, causing water quality degradation and killing fish and other aquatic organisms. For instance, in Oregon, a tractor-trailer filled with hydrochloric acid jackknifed and ran off the road, spilling its contents into the John Day River. In California, derailed train cars carrying herbicides plunged into the nearby Sacramento River, spilling their contents. In Idaho, a tractor-trailer full of industrial chemicals ran off the highway into a tributary of the Salmon River. In Utah, chlorine from a water treatment facility spilled into the Ogden River.

Large spills of oil and other chemicals do not occur frequently, but the results are often catastrophic when they do. Some of the more common types of spills, however, are spilled petroleum products from 55-gallon drums, overflows from sewage pump stations, breaks in sewer lines, breaks in oil and gas lines, and spills of gasoline and antifreeze. These lesser spills also pose a threat to water quality.

Water Control Structures

Dams, levees, and other water control structures alter the natural courses of our rivers and streams. Some control structures turn free-flowing waterways into standing bodies of water. These changes in the natural movement of our waterways can reduce

Bennett Dam, Peace River, British Columbia

water quality in a number of ways. For instance, the temperature of water held behind a dam increases because more water surface is exposed to the sun and less vegetation exists to provide shade. This increase in temperature reduces the amount of oxygen that will dissolve in the water (see Chapter 2) and become available for fish and other aquatic organisms. The water held behind an impoundment usually has less natural turbulence than a moving waterway, also resulting in lower concentrations of dissolved oxygen.

By slowing down the movement of the water, control structures also cause pollutants suspended in the water, such as nutrients and sediments, to settle out of the water and concentrate. This slow-moving, warm water with a high concentration of nutrients provides a good environment for algae. As algae grow and multiply, the water can become green and turbid. Algae can also cause unhealthy shifts in dissolved oxygen and pH, as described earlier.

Control structures can severely alter fish habitat. Waters passing through a dam can become over-saturated with dissolved gases, such as nitrogen, that are harmful to fish and other aquatics. Silt and other materials settling from the water held behind a dam can cover fish spawning beds. Migratory routes can be completely cut off or made more difficult. In the United States, many runs of both Atlantic and Pacific Salmon have been devastated because water control structures have disrupted their migratory routes and habitat.

The loss of natural flood plains and wetlands because of the use of dikes and levees also reduces water quality. Flood plains filter out and remove pollutants as water passes through vegetation and over soil during flooding. These benefits are lost when dikes are placed to prevent water from entering them.

In summary, although water control structures provide benefits to society, such as hydropower, irrigation, and flood control, they also contribute to water pollution. These control structures should be recognized not only for their benefits to society, but also for their costs to the environment.

Summary

The purpose of this chapter was to introduce you to some of the sources of water pollution: stormwater runoff, domestic discharges, industrial discharges, accidental spills, and water control structures. You learned that stormwater runoff is currently the leading cause of water pollution in the United States. In the next chapter, you will discover methods to prevent and control these different sources of pollution.

Additional Reading

Borrelli, P., 1989. *Troubled Waters, Alaska's Rude Awakening to the Price of Oil Development.* The Amicus Journal, a publication of the Natural Resources Defense Council, Summer 1989, Vol. 11, Number 3.

Environmental Protection Agency, 1994. *Clean Water Act Fact Sheets: Controlling Agricultural Sources of Water Pollution.* Office of Water (WH-4105F), May, 1994.

Grossman, E., 2002. *Watershed: The Undamming of America.* Counterpoint, New York, New York.

Haberman, R., 1995. *Dam Fights of the 1990s: Removals.* River Voices, a quarterly publication of the River Network, Volume 5, Number 4/Winter 1995.

Helvarg, D., 2001. *Blue Frontier: Saving America's Living Seas.* W. H. Freeman and Company, New York, New York.

Laws, E. A., 1993. *Aquatic Pollution: an Introductory Text.* John Wiley & Sons, Inc., New York, New York.

Mitchell, J. G., 1996. *Our Polluted Runoff.* National Geographic, Vol. 189, No. 2, February 1996.

National Geographic Society, November 1993. *The Power, Promise, and Turmoil of North America's Fresh Water*, National Geographic Society, Washington, D.C.

National Research Council, 2001. *Upstream, Salmon and Society in the Pacific Northwest.* National Academy Press, Washington, D.C.

Nichols, A. B., 1990. *Prince William Sound Starts to Recover.* Water Environment and Technology, a publication of the Water Pollution Control Federation (now called Water Environment Federation), May 1990.

Peterson, K.C., 2001. *River of Life Channel of Death: Fish and Dams on the Lower Snake.* Oregon State University Press, Corvallis, Oregon.

ReVelle, P., and C. ReVelle, 1988. *The Environment: Issues and Choices for Society, Third Edition.* Jones and Bartlett Publishers, Boston, Massachusetts.

Safina, Carl, 1995. *The World's Imperiled Fish.* Scientific American, Vol. 273, No. 5, November, 1995.

World Commission on Dams, 2000. *Dams and Development, A New Framework for Decision-Making.* Earthscan Publications, Sterling, Virginia.

4. Preventing Water Pollution

Carrabassett River, Maine

How do we prevent water pollution? What methods are available to control pollution from stormwater runoff? What about municipal and industrial discharges, or accidental spills—how do we keep them from damaging our waterways? Fortunately, many techniques are available to eliminate and reduce water pollution from these and other sources.

Environmental professionals have developed many different ways of preventing and controlling water pollution. Some of these methods are as simple as managing and storing materials properly to prevent spills from occurring. Others involve more complex treatment processes.

Lake Superior tributary, Wisconsin

Prevention

The old adage "an ounce of prevention is worth a pound of cure" applies to water pollution control. It is easier and cheaper to keep pollutants from entering our waters than it is to remove them. Some of the best opportunities available for preventing water pollution involve reducing, reusing, and recycling.

Reducing means using less of an item or creating less waste. When we follow water conservation measures at home and in industry, we allow clean water to stay free of contaminants by leaving it at its source. We can also help prevent water pollution by reducing the generation of waste materials and wastewater. For example, manufacturers of laminated wooden beams minimize waste by reducing the amount of glue wasted in making the beams. Instead of rinsing off excess glue from the beams with water and creating glue-contaminated wastewater, employees carefully scrape off excess glue and save it for future use. This type of waste reduction provides both environmental and economic benefits.

In fact, almost any technique used to reduce waste helps prevent water pollution. When we reduce our generation of garbage and other refuse, less solid waste ends up in landfills. Less solid waste in landfills provides less opportunity for creating landfill wastewater, called leachate. Leachate is created when groundwater or surface water mixes with refuse in a landfill. In addition, we can improve water quality by reducing our use of harmful chemicals. Using less chlorine bleach in paper making, for example, minimizes the generation of harmful chlorinated organic compounds.

Reusing—using the same item more than once—also helps prevent water pollution. In the example used above, laminated beam manufacturers reuse the excess glue after scraping it off the beams. This procedure results in less wastewater and therefore protects water quality.

Reusing water provides at least three important benefits: it reduces the demand for clean water, reduces the quantity of wastewater requiring treatment, and reduces the amount of treated wastewater discharged back into the environment. Some industries have been particularly effective in reusing water. They reuse process waters for wash down and cleanup rather than using clean water. Some municipalities reuse treated and disinfected wastewater for irrigation.

Recycling—using the same item more than once in the same or alternate forms—is the third conservation practice that helps prevent water pollution. For example, paper recycling helps prevent water pollution by lowering the demand for raw timber, allowing more trees to remain on the mountainside for stabilizing the soil, cooling tributary waters, and otherwise benefiting water quality. Recycled paper is also easier to pulp than timber. It takes less energy, less water, and fewer chemicals to create recycled paper than it does to create paper from raw wood.

We also can recycle other items to help prevent water pollution, including glass, aluminum, oil, metals, and plastics. Using these items in their recycled forms requires fewer virgin resources, disturbs the land less, and generally consumes less energy. Instead of developing a new mine to extract metal deposits, for instance, we can increase our use of recycled metals to meet all or part of the need.

Chapter 8 describes in more detail several methods of reducing, reusing, and recycling in the home to prevent water pollution.

Other simple ways of preventing water pollution include storing materials that could become pollutants away from areas where they may enter the water. Many industries have developed material storage plans that outline where particular materials are to be stored. Materials that could become waterborne pollutants are not stored near stormwater inlets, gutters, ditches, or other areas where they could easily enter the water accidentally.

Industries also prevent water pollution by properly transferring and handling materials such as petroleum products in order to prevent spills. Later sections of this chapter outline spill prevention and cleanup procedures.

Natural Water Pollution Control Processes

Water quality professionals often categorize water pollution control processes as *physical, chemical,* or *biological.* Physical processes rely on the physical separation of pollutants from the water, chemical processes rely on chemical reactions, and biological processes rely on living organisms to break down waste materials.

All techniques used to control water pollution are based on one or more fundamental processes that occur in nature. These natural processes include sedimentation, biodegradation, filtration, and sorption.

Sedimentation describes the process of particles settling in water. It occurs naturally as a result of gravity. As the force of gravity, which pulls everything towards the center of the earth, acts on the particles, they fall to the bottom of the water body. Sedimentation is enhanced if the water body is still, such as in a lake, pond, or basin where the forces of turbulence and mixing found in fast moving waters are absent. Turbulent forces tend to counteract the force of gravity and keep materials suspended in the water.

The Mississippi River provides a good example of sedimentation in nature. Many years ago, prior to construction of locks and dams, the Mississippi flowed freely. Sand and other soil particles that eroded from the banks of the river became suspended in the river as it moved south from what is now Minnesota to Louisiana. As the river entered the Gulf of Mexico near what is now New Orleans, the turbulent forces of the river dissipated. Without the turbulent forces to keep the materials suspended, the sand fell to the bottom of the gulf due to gravity. This sedimentation process, occurring over thousands of years, created the Mississippi Delta—the broad, flat, fan-shaped area at the confluence of the Mississippi River and the Gulf of Mexico.

Water quality professionals use the natural process of sedimentation in controlled settings—at industrial and municipal treatment plants, for instance—to remove suspended pollutants from the water. Sedimentation is a *physical* treatment process.

Natural biodegradation is also a common water pollution control process. As you already know from Chapter 2, biodegradation occurs as microscopic organisms, especially bacteria, break down organic substances into simple carbon and nitrogen compounds and water. The process is called biodegradation because biota are responsible for decomposing, or degrading, the organic substances. Bacteria use the organic materials as food for energy and growth. Compounds generated in the biodegradation process not used by bacteria, such as carbon dioxide, are recycled back into the environment.

Biodegradation is one of the fundamental processes in nature responsible for recycling nutrients, such as carbon, nitrogen, and phosphorus. Communities and industries rely on biodegradation in controlled settings—at municipal and industrial treatment plants—to break down and remove waste materials from their wastewaters. When biodegradation is used in a controlled setting to treat wastewater it is called *biological* treatment.

When you make a pot of coffee, you usually use a coffee filter to keep the solid coffee particles out of the liquid. This process, called filtration, also occurs naturally in the environment. As water passes over the surface of the land, it moves through grass and other vegetation or seeps into the soil. The vegetation and the soil act as natural filters. Materials that are larger than the openings between adjacent pieces of vegetation or grains of soil are filtered out.

Natural filters come in different sizes. For example, a bed of willows is a natural filter capable of removing large debris such as leaves, branches, and twigs from the water. A bed of sand is a natural filter that can remove smaller particles such as small pieces of vegetation and large soil particles.

Communities and industries use filtration in controlled settings for water pollution control. Municipal and industrial treatment plants employ racks, screens, filters, and beds of sand to filter out different sizes of particles from wastewater. Filtration is another example of a *physical* treatment process.

Natural filtration, Roaring River, Oregon

If you spilled water on the kitchen floor, you would probably use a towel, cloth, or sponge to clean it up. These items work well because they absorb the water. Absorption, then, is the process where one substance is taken into another. Some of the materials found in nature, like certain types of soil and vegetation, are good at absorbing pollutants. For instance, soils containing organic materials such as peat moss and decaying vegetation remove some types of organic pollutants by absorbing them.

Another natural process, similar to *ab*sorption, is called *ad*sorption. Adsorption is the process where one material simply adheres to the surface of another material. Absorption and adsorption occur together in nature and it is difficult and usually unnecessary to distinguish the two. When these two processes are combined, the resulting process is simply called sorption.

Communities and industries use the sorption process in their wastewater treatment plants to remove pollutants. Sorption is a combined *physical* and *chemical* treatment process.

Stormwater Treatment

As explained in Chapter 3, nonpoint sources of pollution associated with stormwater runoff are the leading cause of water pollution in the United States today. A variety of methods are available for controlling and preventing stormwater pollution. Some of these stormwater treatment methods are new and some have been around for decades, if not centuries. Some use new applications of the fundamental treatment processes described above—sedimentation, biodegradation, filtration, and sorption—and some rely on new and old erosion control techniques.

Atlantic Ocean near Fire Island National Seashore, New York

For example, some cities use grass-lined swales—also called bioswales—as part of new highway drainage systems to remove stormwater pollutants from highway runoff. Bioswales are simply flat drainage ditches lined with trees, shrubs, grasses, and other vegetation. Runoff from the highway that is collected in gutters and drainage pipes flows into swales built parallel to the roadway. As water flows through grass and vegetation, pollutants sorb onto the surfaces of the soil and vegetation in the swale. In essence, the runoff is "scrubbed" as it flows down the swale. By the time the runoff reaches a nearby creek or stream, many of the stormwater pollutants have been removed.

Water quality engineers also design bioswales to treat runoff from residential, commercial, and industrial properties. Often, swales become part of a site's landscaping. Bioswales are simple and relatively inexpensive, but they serve at least two important functions: they add vegetation to the site, making it more attractive, and they help control stormwater pollution.

Stormwater ponds are another method for removing stormwater pollutants. They function by first reducing the velocity of the stormwater entering the pond, allowing sedimentation to occur. Pollutants are also removed by vegetation and soil in the pond through filtration and sorption, and by bacteria through biodegradation. Stormwater ponds help prevent flooding by detaining stormwater runoff, which also prevents erosion.

People have used erosion and sediment control techniques for many years to conserve soil. Now, we also recognize their value for keeping soil and other pollutants out of our waterways. Erosion and sediment control can be accomplished by preventing stormwater from contacting the soil, reducing the velocity of stormwater, stabilizing the soil, or capturing eroded sediments at the source.

You may have seen some methods of controlling erosion and sediment used in your area. Ten of the most common methods are:

✔ diverting stormwater runoff away from areas where the soil has been disturbed;

✔ planting grass and other vegetation to stabilize the soil and prevent it from being washed away;

✔ placing burlap, coconut fiber, or synthetic mats to stabilize disturbed soils and keep them in place;

✔ adding cobbles and boulders, or *riprap*, to areas where the stormwater velocity is high—at the discharge ends of pipes, for instance—to keep the soils in place and reduce the erosive energy of the water;

✔ adding gravel pads to the entrances of construction sites so soil is not tracked off by vehicles entering and exiting the site;

✔ placing plastic covers over disturbed or stock-piled soils to keep rainwater from contacting the soil and causing erosion;

✔ compacting and texturing disturbed soils carefully with heavy equipment to make them more resistant to erosion;

✔ placing straw bales in ditches to slow down the velocity and reduce the energy of the stormwater;

Silt fences and grass stabilization at a construction site

✔ placing barriers such as black plastic silt fences around disturbed areas to keep eroded soils from washing off the site; and

✔ placing barriers around drains and inlets to keep soils from washing into them.

A combination of these methods works best to control erosion and sediment. For example, at large construction sites where soil is being disturbed extensively, contractors typically use silt fences, straw bales, gravel construction entrances, and inlet barriers. They usually also seed the disturbed areas as soon as possible with grass and other vegetation to stabilize the soil.

Sometimes communities prohibit construction activities during rainy periods to prevent erosion. Other times they allow construction during rainy periods if proper erosion control measures are being used and properly maintained.

Oil/water separators are devices that treat stormwater runoff from parking lots and other areas containing petroleum products. Although these devices come in many different shapes and sizes, they are all based on one fundamental principle: oil is lighter than water. Oil floats to the surface of water, where it can be removed. In some of these separators, stormwater passes through a concrete box containing baffles that remove the floating oil. In others, stormwater flows through a series of plates that help separate the floating oil from the water. Oil/water separators work best for removing concentrated discharges of petroleum products. They are only partially effective in removing petroleum products that have become dispersed throughout the stormwater.

Vegetated filter strips, infiltration galleries, leaf-compost filters, and catch basin filters also remove pollutants from stormwater.

Vegetated filter strips are strips of grass and other vegetation. They function much like bioswales—removing pollutants as water passes over and through them—except that they are flat, sloping strips of land instead of drainage channels.

Infiltration galleries are pond-like areas where stormwater is collected and allowed to infiltrate into the soil. As the stormwater moves down through the soil, it is also treated through sorption and biodegradation. Before using infiltration galleries at a particular site, however, engineers should consider the potential effects on groundwater quality.

Leaf-compost filters function like bioswales, except that they consist of densely packed leaves. The filter removes pollutants as the runoff flows through the leaves and the pollutants sorb and biodegrade.

Catch basin filters are baskets of filter material, such as sand, leaves, and carbon, which are placed under stormwater inlets and grates. Stormwater entering a catch basin must pass through these baskets of material before entering the drainage system.

Perhaps the biggest unknown in the practice of treating stormwater is what to do with the residuals retained in these various stormwater treatment processes. Communities must properly manage these materials to keep the treatment processes functioning well and to protect the environment. Some of the remaining pollutants may be relatively nontoxic and they can be used as fertilizers or soil amendments. Others may require treatment prior to disposal. Researchers currently are investigating different ways of managing stormwater treatment residuals.

Domestic Treatment

Septic Systems

As discussed in Chapter 3, people living in rural areas commonly treat and dispose of their sewage by using septic systems. With these systems, household wastewater flows from home plumbing into a septic tank and drain field buried in the yard. Most of the sewage solids are retained in the septic tank, where they naturally decompose through biodegradation. Every five to ten years, nonbiodegradable solids such as sand and grit must be removed by having a septic contractor pump the tank. The liquid effluent flows out of the septic tank into a series of perforated drainpipes called a drain field, and then out of holes in the drainpipes into the underlying soil. When the system is functioning properly, effluent from the tank receives further treatment through filtration, sorption, and biodegradation as it percolates down through the soil. Septic systems are sometimes referred to as on-site systems because they are usually located on the property where the wastewater is generated.

People sometimes use another on-site system called a mound system, which is similar to a septic system except that the drain field is located in a mound of earth built on top of the natural surface of the land. A pump conveys effluent from the septic tank to the mound of earth where treatment occurs as the liquid percolates down through the soil in the mound. People use mound systems in areas where using a regular drain field would cause water quality or human health concerns, such as in areas with clay soils or shallow water tables.

Sewage Treatment Plants

Domestic wastewater from most urban areas flows from home plumbing into sewer pipes located under the street and then to a community's sewage treatment plant. Most communities treat their sewage using a common, five-step process.

First, the sewage flows into an area referred to as the headworks. At the headworks, the sewage passes through a coarse screen that removes miscellaneous solids that get into the sewer system, such as paper products, cloth, and plastics. Solids making it through the coarse screen are sometimes chopped and ground into smaller particles by a device known as a comminutor. The wastewater then passes through an area called a grit chamber where sand, gravel, cinders, and other similar solids are removed. At the headworks, a specially designed channel called a Parshall flume is often used to measure the flow rate of the incoming sewage. Alternatively, the flow rate is measured with a flume at the end of the treatment process.

Second, the sewage flows into a large tank called a primary sedimentation basin or primary clarifier. These tanks are often circular and may be thirty or more feet across and fifteen or more feet deep. Here, the larger and heavier solids that remain in the sewage settle out of the solution due to gravity.

Secondary clarifier

Third, the sewage—minus the removed solids—flows from the primary clarifier into another basin, where large paddles or brushes add air by vigorously mixing the wastewater. Alternatively, perforated air pipes on the bottom of the basin add air similar to the way a bubbler adds air to a fish tank. This aeration basin provides the oxygen needed by bacteria for biodegrading the organic materials in the sewage. Treatment through biodegradation in an aeration basin is simply the process of converting the organic waste materials in the sewage into new bacterial cells, carbon dioxide, and water.

Next, the outflow from the aeration basin, consisting mainly of water and bacterial cells, flows into a large tank called a secondary clarifier. Here, the bacterial cells and other remaining solids settle to the bottom of the clarifier and a mechanical sweeping device continually removes them.

Some of the bacterial solids removed from the secondary clarifier are returned to the aeration basin. These solids, called activated sludge, help maintain an active mass of microorganisms in the aeration basin to promote further biodegradation. When bacterial solids are recycled in this manner, the entire treatment system is usually called an activated sludge plant.

Finally, the water flows out of the secondary clarifier to a chamber or basin for disinfection. Most communities disinfect their wastewater by adding chlorine and mixing it thoroughly with the effluent. Sufficient time is provided in the chlorine contact chamber for the chlorine to kill the harmful bacteria and viruses left in the wastewater.

Other methods of disinfecting wastewater are available, such as adding ozone or passing the treated wastewater through ultraviolet light. These methods of disinfection help prevent the toxic effect of chlorine on fish and other aquatic organisms in the receiving water.

After communities treat their wastewater with this five-step process, they commonly discharge it into the nearest river or stream. Alternatively, they dispose of the treated wastewater on land using irrigation techniques. The treated wastewater is called effluent.

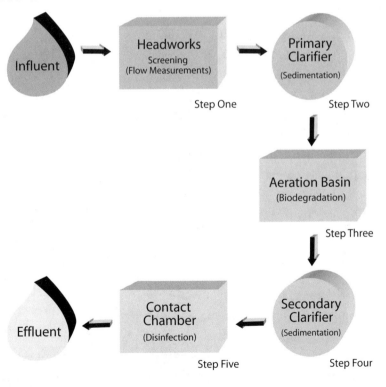

Five common steps for treating sewage.

The solid materials removed from sewage—mainly in the primary and secondary clarifiers—are called biosolids, or sludge. Communities must properly manage these solids to protect the environment and public health. Biosolids management includes collecting, digesting, and stabilizing the solids through biodegradation and pH adjustment, and dewatering and drying the solids before disposal. Communities dispose of their sludge by sending it to a landfill, burning it, or using it as fertilizer on certain types of land. The practice of using the biosolids as fertilizer is often the best disposal alternative, because it allows the nutrients in the biosolids to be used by vegetation growing on the disposal site. Biosolids application must be done carefully, however, at rates that allow the vegetation to use the nutrients properly. These rates are referred to as agricultural rates or agronomic rates.

Other methods of treating municipal wastewater are adaptations of this basic five-step process. For instance, wastewater treatment lagoons also rely on this five-step process, but primary sedimentation, aeration, and secondary sedimentation occur in the lagoon instead of in separate basins. Sometimes a rock filter replaces the aeration basin. In this type of system, wastewater travels from the primary clarifier to a basin of round river rocks, usually five to ten feet deep. Bacteria grow on these rocks and biodegrade the wastewater as it trickles down through the filter. Some treatment plants use other types of filter media, such as wooden slats or plastic materials, in place of the river rocks. This type of treatment system is called a trickling filter.

Package treatment plants, which are systems where all processes are contained in one package, also use the five-step process, but the different steps occur in separate chambers of one large tank. Small developments like trailer parks and resorts sometimes use these package plants if they are located in areas without a community sewage treatment system.

Some communities even rely on wetlands to treat their domestic wastewater. They turn upland areas into wetland treatment systems by excavating ponds and swales, planting vegetation, and supplying wastewater. Engineers, scientists, and the public are paying more attention to these constructed wetlands recently because they sometimes offer an aesthetic alternative to conventional mechanical treatment systems. *Constructed* wetlands remove pollutants through the processes of sedimentation, filtration, sorption, and biodegradation as wastewater flows through the wetland. *Natural* wetlands can "polish" domestic wastewater—in other words, provide additional treatment—following a specified level of pretreatment using other processes. However, natural wetlands may not always be a desirable alternative for treating or polishing domestic wastewater because of the potential harm to the vegetation and aquatic organisms in the wetland.

Industrial Treatment

As mentioned previously, many industries use water in their production processes. To protect the environment, they must remove the pollutants added during production prior to discharging the water. Like municipalities, industries commonly use the fundamental processes of sedimentation, biodegradation, filtration, and sorption to remove pollutants, along with the processes of precipitation and neutralization.

Unlike municipal treatment techniques, however, which are similar from one community to the next, industrial treatment techniques are industry-specific. Treatment processes vary even within a single industry, depending on the nature of the pollutants found in the wastewater of a particular manufacturer or factory. For example, industries generating wastewater with high concentrations of solids and organics use physical processes to remove solids and biological processes to remove organics. Potato processors—one of many types of food processors—first wash the potatoes to remove soil and then rinse the peeled potatoes to remove starch. The water used in these processes acquires soil particles and starchy, organic potato wastes. This type of wastewater is similar to domestic wastewater, though it has a much higher concentration of solids and organics. Because the two wastewaters are similar, the treatment processes employed are similar. Generally, potato processors treat their wastewater to remove primary solids, such as soil particles and potato peels, by passing the wastewater through screens and allowing it to stand in sedimentation basins. They then remove dissolved organics by allowing bacteria to biodegrade the organic materials and convert them into new bacterial cells. Finally, they remove the secondary solids, consisting mainly of bacterial cells, by allowing them to settle in a secondary sedimentation basin.

Even after this level of treatment, potato-processing waters still have relatively high concentrations of pollutants. They usually are not clean enough to be discharged into surface waters. Instead, the industry typically discharges the treated wastewater onto land, where further treatment occurs as the wastewater percolates through the soil. As you may recall, this method of further treating and disposing of wastewater by applying it to land is also used for municipal wastewater and biosolids. It provides crops with nutrients from waste materials. The industry must apply these wastewaters at agronomic rates, however, to allow crops to fully use the nutrients. Otherwise, the nutrients will leach into the groundwater, possibly causing problems like nitrate contamination.

Other industries with wastewaters containing solids and organics use similar treatment processes. For example, pulp and paper mills use screening and sedimentation to remove solids, such as wood chips and fibers, and biological treatment to remove organics, such as the carbon compounds found in the wood.

A variety of industries use *chemical* treatment to remove pollutants by employing treatment processes that rely on chemical reactions. For instance, industries manufacturing metals, or using metal components in their products, often generate wastewater with high concentrations of dissolved metals. To remove dissolved metals, they add alkaline chemicals such as sodium hydroxide to the wastewater to increase its pH. When the pH rises, many of the dissolved metals come together to form solids. The chemical reaction responsible for causing the dissolved metals to combine and form solids is called a precipitation reaction. Once formed, gravity causes these solids, which are called precipitates, to fall from the solution. This process, which occurs in a sedimentation basin, effectively removes many of the dissolved metals from the wastewater. After the solids settle, the pH of the wastewater remains high and industries then rely on another treatment process called neutralization to bring the water back to a neutral pH range.

Recall from Chapter 2 that liquids with a pH of 7.0 are neutral, liquids with a pH less than 7.0 are acidic, and liquids with a pH greater than 7.0 are basic. Most natural waters have a pH ranging from about 6.0 to 8.5. When wastewater is either too acidic or too basic, it is harmful to aquatic organisms and must be neutralized prior to discharge in order to protect the environment.

Acidic and basic substances neutralize each other. Industries neutralize acidic wastewater by adding bases and basic wastewater by adding acids. In the example used above, the wastewater remained basic after the metals were removed through the precipitation reaction. A metal manufacturer would neutralize this basic wastewater by adding an acid with a low pH, such as hydrochloric acid, and bring the pH to approximately 7.0. Manufacturers of rare metals, electronic equipment, chemicals, and pharmaceuticals are just a few of the many industries that use neutralization to treat their acidic or basic wastewaters.

Spill Prevention and Cleanup

Industrial manufacturing plants have relatively constant waste streams, especially if they are making a single product. Once the pollutants in the waste streams are identified and the volume of the wastewater being generated is known, industries can develop good treatment processes to prevent and control water pollution. It is more difficult to prevent and control less predictable sources of water pollution, such as accidental spills.

Spilled petroleum products and other chemicals can severely damage the water environment. Those people responsible for petroleum products and other chemicals can prevent spills and reduce their impact by following good planning, design, operation, and maintenance procedures.

For example, storing bulk liquids like chemicals and petroleum products warrants special attention. These materials should be stored in strong vessels and inspected regularly. Moreover, bulk liquids should have primary containment and secondary containment, in case the first method fails.

Secondary containment is commonly provided by placing the primary containment vessels—steel tanks, for instance—into concrete bunkers. These bunkers usually consist of a concrete floor and walls, but no roof. They are usually designed to hold one and one-half times the volume of the largest tank placed in the bunker, so that if a tank breaks, the material stored in it will be entirely contained within the concrete bunker, where it can be recovered.

Vessels used to transport bulk liquids by rail and barge need secondary containment also. Large barges transporting bulk liquids, for example, usually have an additional inner hull in case the outer one breaks. Double-walled steel tanks are often used to transport bulk liquids by rail.

We can also prevent spills by designing highways and railroads properly. Any design element that reduces the opportunity for accidents also reduces the likelihood of accidental spills. For instance, engineers typically design highways and railroads so that grades are not too steep and curves are not too sharp. They also design and place traffic signs and signals appropriately. Highway and railroad crews are responsible for maintaining our highways and railroads to prevent accidents and subsequent spills.

In the United States, federal law requires Spill Prevention, Control, and Countermeasure (SPCC) plans for sites where petroleum products could potentially spill and contaminate nearby waters. SPCC plans include an inventory of all petroleum products showing where they are located and the type of storage vessel used. The plans include topographic information showing the direction materials would travel if spilled and descriptions of local soil and groundwater conditions, plus drawings of any natural or constructed drainage system the spilled materials could enter. The plans usually have figures or photos showing the nearest surface water and how spilled materials could enter it. They outline procedures for properly loading and unloading materials to prevent and control spills.

Storage tanks with secondary containment

SPCC plans also include procedures for responding to and cleaning up spills. The plans include a list of emergency contacts, such as qualified employees, a fire marshal, and a county sheriff, and their phone numbers. They contain an inventory of the type and location of emergency response materials such as sorbent cloth, drain plugs, saw dust, protective clothing, gloves, booms, and other materials.

Companies can plan ahead to prevent spills by developing and implementing SPCC plans. Potential problem areas can be identified and site-specific safeguards developed to prevent spills from occurring. Without planning ahead, employees are left to invent their own procedures during an emergency, with whatever materials are on hand.

Summary

This chapter outlined ways of preventing water pollution by utilizing the natural processes of sedimentation, biodegradation, filtration, and sorption. It described how municipalities and industries use these processes to remove pollutants in controlled settings at treatment plants. You learned methods to reduce water pollution caused by stormwater runoff.

Many of the methods used to prevent and control water pollution were developed in response to the laws and regulations enacted to protect our water resources. These regulations have important and far-reaching implications. Water pollution control, then, is not only about science and technology. It is also about the many rules and regulations governing water quality protection, which is the topic of the next chapter.

Additional Reading

Eckenfelder, W. W., 1989. *Industrial Water Pollution Control, Second Edition*. McGraw-Hill, Inc., New York, New York.

Environmental Protection Agency, 1992. *Storm Water Management for Industrial Activities: Developing Pollution Prevention Plans and Best Management Practices*. Office of Water (WH-547), September 1992, EPA 832-R-92-006.

Ferguson, B.K., 1998. *Introduction to Stormwater: Concept, Purpose, Design*. John Wiley & Sons, Inc., New York, New York.

Franzini, J.B., Freyberg, D.L., and G. Tchobanoglous, 1992. *Water Resources and Environmental Engineering*. McGraw-Hill, Inc., New York, New York.

Gordon, W., 1984. *A Citizen's Handbook on Groundwater Protection*. Natural Resources Defense Council, New York, New York.

Gore, J. A., 1985. *The Restoration of Rivers and Streams: Theories and Experience*. Butterworth Publishers, Stoneham, Massachusetts.

Herricks, E.E. (Editor), 1995. *Stormwater Runoff and Receiving Systems, Impact, Monitoring, and Assessment*. CRC Press, Inc., Boca Raton, Florida.

Masters, G.M., 1997. *Introduction to Environmental Engineering and Science, Second Edition*. Prentice Hall, Inc., Upper Saddle River, New Jersey.

Metcalf and Eddy, Inc., 1991. *Wastewater Engineering: Treatment, Disposal, Reuse, Third Edition*. Revised by G. Tchobanaglous and F.L. Burton. McGraw-Hill Book Company, New York, New York.

Novotny, V., and H. Olem, 1994. *Water Quality: Prevention, Identification, and Management of Diffuse Pollution*. John Wiley & Sons, Inc., New York, New York.

Tchobanaglous, G. and E. D. Schroeder, 1985. *Water Quality: Characteristics, Modeling, Modification*. Addison-Wesley Publishing Company, Menlo Park, California.

Walesh, S. G., 1989. *Urban Surface Water Management*. John Wiley and Sons, Inc., New York, New York.

Washington Department of Ecology, 1992. *Stormwater Management Manual for the Puget Sound Basin*. Washington Department of Ecology, Olympia, Washington.

5. Water Quality Regulations

Baker River, New Hampshire

What rules and regulations exist for protecting our water resources? Do we have specific rules for rivers, lakes, or wetlands? What about drinking water—what regulations help assure us it is clean and safe? Many rules and regulations have been adopted in the United States at the federal, state, and local level to protect water quality. These rules and regulations continue to evolve as we learn from our experiences.

This chapter summarizes the water quality regulations currently in place at the federal, state, and local levels. These rules are fundamental to our understanding of water quality and water pollution control in the United States. By becoming informed about the regulations protecting our water, we can track progress in meeting them. We can thoughtfully consider changes, voice our opinions, and work with our elected officials to ensure that their policy decisions help protect our water resources. These rules and regulations dictate many of the procedures used for water quality protection.

Federal Regulations

The United States Congress has enacted many important federal regulations to protect water quality. These regulations have helped control water pollution and protect our drinking water. They also have helped us to clean up contaminated waters and protect waters that provide important aquatic habitat to species whose survival is in question.

Clean Water Act

The Federal Water Pollution Control Act, commonly known as the Clean Water Act (CWA), is the cornerstone of water quality legislation in the United States. It is not the result of a single piece of legislation. Rather, the current Act is a combination of federal water pollution control policies developed over many years. Its legislative history goes back over one hundred years to the

Humpback whales near
Sitka, Alaska

Rivers and Harbors Act of 1899. The United States Congress brought together much of this historic water quality legislation in 1972 when they created Public Law 92-500, now simply called the Clean Water Act.

The Clean Water Act consists of five separate parts, called Titles. Title I is the introductory section, which declares the goals and policies of the Act. According to Title I:

The objective of this Act is to restore and maintain the chemical, physical, and biological integrity of the Nation's waters.

The Clean Water Act also establishes the goal of making the nation's waters fishable and swimmable. Title I also includes descriptions of research and other related programs.

Title II includes a description of the grants program for constructing publicly owned treatment works. This program was responsible for providing funding to construct many of the municipally owned sewage treatment plants in the United States. It provided much of the financial incentive for many of the major sewerage projects constructed from 1972 to 1987. In 1987, the Clean Water Act was amended to phase out the grant program, which was replaced with a revolving fund, low-interest loan program. The loan program is being administered by individual states that receive federal matching funds. Title II also includes a description of the river basin planning program.

Title III of the Clean Water Act is the section on water quality standards and enforcement measures. Standards are the reference values by which we judge the quality of our waters. Individual state water quality agencies typically develop these standards and submit them to the Environmental Protection Agency for review and approval. Because of their importance, water quality standards appear in more detail under the heading of **State Regulations** later in this chapter. Title III also includes a description of programs for developing effluent limitations, reviewing water quality conditions, preventing the discharge of oil and hazardous substances, and maintaining clean lakes.

Section 305(b) of Title III includes the procedures for reviewing or inventorying water quality conditions. Every two years, State agencies inventory the quality of waters in their state and submit a summary report to the Environmental Protection Agency. This 305(b) report is a valuable reference for citizens who are concerned about the status of water quality in their state. It includes an inventory of polluted waters and the suspected sources of contamination. It also describes the efforts being made to improve the quality of these waters. You can usually obtain copies by contacting the agency responsible for water quality control in your state, or by visiting their web site.

Many of the policies and procedures for preventing oil pollution in United States' waters are outlined under Section 311 of Title III. According to Section 311,

> *The Congress hereby declares that it is the policy of the United States that there should be no discharges of oil or hazardous substances into or upon the navigable waters of the United States, adjoining shorelines, or into or upon the waters of the contiguous zone.*

Section 311 requires a National Contingency Plan to be developed for the removal of spilled oil and hazardous substances. The purpose of the plan is to provide an efficient and effective course of action for oil-spill emergencies in order to minimize damage to the environment.

Another program outlined under Section 311 requires owners of sites containing petroleum products to prepare and implement spill prevention, control, and countermeasure plans for these sites. As discussed in the last chapter, SPCC plans outline the procedures to be followed to prevent spills from occurring and to respond effectively if they do. Those people responsible for sites where petroleum products could reasonably be expected to contaminate waters if spilled must prepare these spill contingency plans. (See **Spill Prevention and Cleanup** in Chapter 4.)

Title IV of the Clean Water Act contains programs for water quality permits and licenses, including three you may have heard about: 1) the National Pollutant Discharge Elimination System (NPDES) permitting program, 2) the dredge and fill permitting program, and 3) the Water Quality Certification program. (See **State Regulations** later in this chapter.)

Finally, Title V includes the other general provisions of the Act. These provisions include administrative procedures, definitions, and methods for procurement. Section 505 describes the procedures that an individual citizen can take to file a civil suit against any entity, including the government, for violating terms of the Clean Water Act.

Sometimes regulators and other professionals refer to water quality programs by their numbers and it may seem intimidating. However, these numbers simply refer to the sections of the Clean Water Act where the programs are defined. For example, the 305(b) report is defined in Section 305(b) of the Act, the Section 404 rules for fill and removal of soil in wetlands are defined in Section 404, the 401 Certification Process is described in Section 401, and so forth.

Safe Drinking Water Act

The Safe Drinking Water Act (SDWA) is the key piece of legislation that protects our drinking water. Congress originally passed the SDWA in 1974 to protect public health by keeping our drinking water free from contamination, and it has been amended several times over the years.

The Safe Drinking Water Act defines the maximum concentrations of contaminants allowed in our drinking water. It defines the maximum contaminant levels (MCLs) for inorganic substances, organic substances, and microorganisms. For example, the MCL for nitrate, an inorganic substance, is ten parts per million. The MCL for benzene, an organic substance, is 0.005 parts per million. The MCL for coliform bacteria allows only one water sample per month to test positive for total coliforms when fewer than forty samples are taken per month. Chapter 7 contains a list of the MCLs for other drinking water contaminants.

Salmon Street Springs, Portland, Oregon

The Environmental Protection Agency sets these contaminant levels after reviewing the findings of scientific studies and evaluating public comments. They consider water to be safe to drink when the concentrations of contaminants are below the adopted MCLs. The EPA routinely reviews MCLs in light of ongoing research. They present any proposed changes or additions to MCLs to the scientific community and the public before adopting new standards.

One of the important additions included in the 1986 amendments to the SDWA was a strict schedule for regulating additional contaminants. This schedule required the EPA to increase the number of regulated contaminants from 23 in 1986 to 112 in 1995.

The SDWA requires all communities to adequately test their drinking water for the regulated contaminants. It specifies which contaminants must be tested for and defines the required frequency, the number of samples to be taken, and the specific techniques acceptable for conducting the analyses. Only approved laboratories are allowed to perform most of the analyses required by the SDWA. Individual state health departments are usually responsible for the laboratory approval process.

The Safe Drinking Water Act regulates surface and groundwater sources of drinking water differently. For example, surface water sources generally have more stringent treatment requirements than groundwater sources. Water providers must typically both filter and disinfect surface water sources before delivering the drinking water to the consumer. They may only need to disinfect some groundwater sources. This approach has been adopted because groundwater sources are usually less susceptible to contamination because they are below the land surface, where most pollution occurs.

Because of the concern over the health effects of lead and copper, the Safe Drinking Water Act requires communities to review their water distribution systems to evaluate pipe materials and report this information to the agency responsible for drinking water in their state. Lead contamination can originate from pipes, from solder and interior linings used on the distribution mains, or from home plumbing. Copper contamination can originate from alloys used in distribution mains or home plumbing. The 1986 amendments to the SDWA prohibit the use of lead pipes, solder, or flux in drinking water systems. State and local plumbing codes enforce this prohibition. Chapter 7 contains more information about copper and lead in drinking water.

Drinking water providers must notify the public if the maximum contaminant levels established in the SDWA are exceeded in a community's drinking water supply. The providers must publish the violation in a daily newspaper circulated in the area served by the water system. They also must report the violation by direct mail—in their monthly water bill, for example. If the violation poses an immediate threat to public health, they must announce the problem by radio and television. The water provider must also notify the public of the action being taken and the anticipated schedule for addressing the problem.

National Environmental Policy Act

The United States Congress passed the National Environmental Policy Act (NEPA) in 1969. It was one of the first pieces of legislation in the United States to specifically address protection of the whole environment—air, land, and water—and the organisms living in them, including humans. NEPA is also directed at protecting the socioeconomic environment, and cultural and historic resources.

The purpose of NEPA is to evaluate the environmental impacts of all activities sponsored by the federal government. According to NEPA, the proponents of projects that involve federal funding or other federal interest must complete environmental reviews and prepare environmental documents to evaluate their proposed action and reasonable alternatives, and the environmental consequences of each.

These environmental documents take the form of either an Environmental Assessment (EA) or an Environmental Impact Statement (EIS), depending on the expected level of environmental effects. NEPA requires an EA when the effects of the proposed action are not expected to be significant, according to the federal definition. NEPA requires an EIS when the impacts are expected to be large, far-reaching, or significant in other ways.

The National Environmental Policy Act opens the door on the federal decision-making process. It provides the means for involving state and local agencies and the public in all federal decisions affecting the environment. Decisions at the state and local level may also be subject to NEPA. For example, if a community secures federal money for an activity with environmental concerns, such as highway construction, that project is subject to a NEPA review.

One of the common misconceptions about Environmental Impact Statements is the role they play in stopping a project. An EIS will not necessarily stop an unwise project, even though one of the alternatives evaluated in the EIS is the "No-Action" Alternative. The EIS process provides the public with an opportunity to participate in the environmental review. It ensures that the project proponent reviews the environmental

Palisades Reservoir, Wyoming

consequences of a proposed action and alternatives and makes this information available to the public and federal decision-makers. The NEPA process provides no assurance, however, that the decision-makers will choose wisely. It ensures only that federal activities affecting the environment proceed in an informed and organized way through an open public process.

These environmental documents have at least five basic components:

✔ a description of the affected environment,

✔ a description of alternatives being considered, including doing nothing (the No-Action Alternative),

✔ an evaluation of the environmental impacts of each alternative,

✔ a description of ways to minimize or reduce negative impacts from each alternative (referred to as mitigation), and

✔ a section identifying the recommended alternative.

Although individual federal agencies develop their own methods for implementing NEPA, within specific guidelines, the basic procedures are essentially the same. The NEPA process consists of preparing the environmental document, issuing a draft document for review and comment, revising the document if necessary, issuing a revised final document for review and comment, and rendering a decision based on the final environmental document and review comments.

Endangered Species Act

Congress passed the Endangered Species Act (ESA) in 1973. It was the first powerful piece of legislation to recognize the inherent value of the many different life forms in the environment and establish regulations to protect them. Originally, the ESA listed only 109 species of plants and animals as threatened or endangered. Today, the list has grown to about 900—or 1,400 if species found internationally are included. The ESA requires special protection measures for species listed as threatened or endangered, and their habitat. The United States Fish and Wildlife Service and the National Marine Fisheries Service are the two federal agencies chiefly responsible for implementing the ESA.

Two success stories provide examples of the ESA's value in saving threatened and endangered species. In the 1960s, only 400 breeding pairs of bald eagles existed in the United States' lower 48 states. The ESA helped force a ban on the use of the herbicide DDT, which had been found to be responsible for weakening egg shells and causing eagle mortality. The number of breeding pairs has now increased to 4,000 or more. The bald eagle moved from the threatened to the endangered list in June of 1994. Likewise, the California gray whale is no longer a threatened or endangered species. Federal agencies removed the California gray whale from the threatened and endangered list completely in June of 1994. The number of gray whales has grown from a historic low of a few thousand to about 24,000 today, due in part to protection provided under the ESA.

Although you may not think of it in this way initially, the Endangered Species Act is a form of water quality legislation. In fact, the ESA's list of threatened and endangered species contains more aquatic plants and animals than land-dwelling ones. Since all aquatic species need clean water for survival, the ESA provides a regulatory basis for protecting water quality where these species are threatened or endangered.

When other environmental laws, such as the Clean Water Act and the National Environmental Policy Act, are not sufficient to protect both the water and the species dependent on it, then the ESA becomes an important water quality regulation. The plight of salmon in the Pacific Northwest is a good example of the ESA acting to protect aquatic species and water quality. The salmon populations in the Northwest have declined drastically over the past twenty years, causing several species to be placed on the ESA's endangered and threatened species list. For instance, the Snake River sockeye salmon and spring/summer Chinook salmon were listed as endangered in 1991 and 1994, respectively. To help these endangered species, the National Marine Fisheries Service has developed a recovery plan, as required by the ESA. The recovery plan includes provisions to improve fish habitat by protecting watersheds, stream corridors, and water quality.

Hazardous Waste Regulations

The United States Congress has enacted several laws to help prevent hazardous wastes from damaging the environment and harming public health. Two of the most focused laws are the Resource Conservation and Recovery Act (RCRA) and the Comprehensive Environmental Response, Compensation and Liability Act (CERCLA), also known as the Superfund Act.

Congress enacted RCRA in 1976. Its primary purpose is to ensure that hazardous wastes are managed properly from the time they are generated until they are ultimately disposed of or destroyed. This lifetime management approach is often referred to as the cradle-to-grave concept.

According to RCRA, wastes are hazardous if EPA evaluates them and lists them as such, or if they are ignitable, corrosive, or reactive to certain test substances. Wastes are also hazardous according to RCRA if they cause toxicity in a special test called the Toxicity Characteristic Leaching Procedure (TCLP).

Congress modified RCRA in 1984 with the Hazardous and Solid Waste Amendment, giving EPA and the states additional authority to regulate the disposal practices of hazardous waste generators.

The amendment established three distinct categories of hazardous waste generators:

✔ Large Quantity Generators, which are those facilities generating more than 2.2 pounds (one kilogram) of acute hazardous waste or more than 2,200 pounds (about five 55-gallon drums) of any hazardous waste;

✔ Small Quantity Generators, defined as facilities generating between 220 and 2,200 pounds of hazardous waste; and

✔ Conditionally Exempt Small Quantity Generators, which generate less than 220 pounds of hazardous waste.

RCRA regulations require hazardous waste generators to properly inventory, label, store, transport, and dispose of hazardous wastes and implement methods to minimize the creation of wastes. Facilities that are conditionally exempt under RCRA are often regulated under other programs, such as state regulations for solid waste.

Although RCRA has been successful in helping to manage the generation and proper disposal of hazardous wastes, it has not been effective in cleaning up abandoned or uncontrolled hazardous waste disposal sites. You may have heard about Love Canal, a site in New York State that received extensive press coverage during the 1970s. At Love Canal, a developer built new homes in an area contaminated with hazardous wastes that had been buried twenty-five years earlier. This abandoned disposal site was responsible for contaminating the area's streams, soil, and groundwater. Some residents, fearful of how the hazardous wastes would affect their health, moved away from the area and filed lawsuits. Authorities had to evacuate others for their own safety.

In 1980, the United States Congress enacted Superfund in response to the problems caused by abandoned hazardous waste disposal sites like Love Canal. The Superfund regulations outline a program for discovering abandoned or uncontrolled sites, evaluating the levels and types of contamination, and cleaning up the sites. They also outline an extensive environmental liability program to hold those responsible for the contamination accountable for damages.

Under Superfund, the Environmental Protection Agency developed the National Priority List, which targeted more than 1,200 sites across the country for evaluation and cleanup. Some of these sites are now clean and others are not.

Superfund has been less effective than originally envisioned for two primary reasons. First, it is extremely expensive to clean up these abandoned disposal sites. It often costs several million dollars to evaluate the contamination and develop alternatives for remedial action, and then several million more to actually clean up a single site. Second, the regulators and the parties responsible for the contaminated sites have found it difficult to agree on how clean the sites need to be in the end.

Although RCRA and Superfund are both land- and public-health-based regulations, they play an important role in protecting water quality. RCRA helps to prevent waters from being contaminated by hazardous substances in the first place, and Superfund helps to clean up sites—including surface and ground waters—that have become contaminated.

State Regulations

In many cases, federal agencies have delegated responsibility for implementation of federal water quality regulations to the states. For example, state agencies often administer at least three federal programs established under the Clean Water Act: the National Pollutant Discharge Elimination System program, the water quality standards program, and the water quality certification program. Generally, both state and federal agencies participate in implementing wetland fill and removal regulations.

National Pollutant Discharge Elimination System

Section 402 of the Clean Water Act authorizes the National Pollutant Discharge Elimination System (NPDES) permitting program. Although federally mandated, the EPA typically delegates the program to individual state water quality agencies for implementation. States use the NPDES program to regulate the discharge of industrial wastewater, municipal wastewater, and

Municipal sewage treatment plant, Coquille Bay, Oregon

stormwater. NPDES permits are required for almost all industrial and municipal discharges and some stormwater discharges.

NPDES permits consist of four parts, commonly called schedules, and a set of General Conditions.

Schedule A of the permit outlines the discharge limitations. For instance, municipal sewage treatment plants typically cannot discharge more than thirty parts per million of biochemical oxygen demand (BOD) or thirty parts per million of total suspended solids (TSS). These are referred to as the 30/30 Discharge Limitations. They define the level of treatment required by municipalities prior to discharging treated wastewater.

Schedule B of the permit defines the monitoring requirements, detailing the frequency of sampling and the type of analyses to be performed. Schedule B may require a municipality to measure temperature and pH on a daily basis and BOD and TSS on a weekly basis.

Schedule C includes the compliance conditions and schedules to be met by the permitted facility. For example, treatment plant operators may need to complete a special study to improve the performance of the treatment plant. Schedule C would outline a mandatory program and schedule for performing this study.

Special conditions are outlined in Schedule D. These conditions may include special requirements for monitoring biota (a process called biomonitoring), training staff, and operating and maintaining the facilities.

The final elements of the permit are outlined in the General Conditions, which includes specific procedures for managing data and records, definitions of terms, and penalties for violations of the permit.

Dischargers must renew their NPDES permits every five years. During the renewal period, staff from the state agency issuing the permit completes an evaluation report and drafts a new permit. Individuals have the greatest opportunity to participate in the permitting process at this time. Once the agency drafts the new permit and the applicant reviews it, the agency releases it for public review and comment. Prior to issuing the permit, agency staff must evaluate comments received during the public review period. Often, they schedule a public hearing to allow individuals to express their opinions about the permit.

Permit holders that do not meet the terms of the permit are subject to fines and penalties, usually based upon the degree and frequency of the violation. State agencies can also revoke permits. Under the Clean Water Act, the permit holder may discharge properly treated wastewater as a privilege, not a right. State and federal regulators can deny this privilege with just cause.

Water Quality Standards

Water quality standards are the numeric values or statements that define the acceptable characteristics of our waters and they give us a frame of reference for protecting their quality. For example, to prevent our waters from becoming too acidic or alkaline, states have adopted water quality standards for pH. Although standards may vary slightly from one state to another, a typical standard would require the pH of a particular water body to be within the range of about 6.0 to 8.5—neither too acidic nor too basic. Most states have adopted water quality standards for temperature, dissolved oxygen, turbidity, bacteria, solids, and toxic substances. Most standards are numeric values, but some are simply statements called narrative standards.

Although federally mandated by Section 303 of the Clean Water Act, the process of adopting standards is performed by individual state agencies. Presumably, state agencies are in a better position to set state standards than the federal government, since they know more about the quality and uses of waters in their state.

When establishing water quality standards, state agencies must consider how water bodies are used. Regulators use the term *beneficial use* to describe the kind of activities that a particular water body is targeted for, which helps dictate the desired quality of that water body. Beneficial uses include the following: domestic and industrial water supply, water contact recreation such as swimming or wading, resident fish and aquatic life, irrigation, fishing, boating, anadromous fish passage, aesthetic quality, and hydroelectric power. For instance, if fish and aquatic life are recognized as beneficial uses in a particular water body, the state must set the water quality standards in that water body at levels that support fish and aquatic life.

The Environmental Protection Agency assists individual states in developing water quality standards, as specified in the Clean Water Act. EPA publishes Quality Criteria for Water, a document

Windsurfers, Columbia River, Oregon

that describes the recommended values for specific water quality standards based on scientific studies. EPA first published Quality Criteria for Water in 1976 and now publishes it about every ten years. The 1976 version is often called the "Red Book" and the 1986 version is called the "Gold Book" because of the colors of their covers.

Municipal and industrial dischargers must not cause water quality standards to be violated in the streams receiving their discharge. They must properly treat their wastewater prior to discharge to achieve this end. In some states, dischargers are allowed a small area in the stream for mixing of effluent prior to meeting water quality standards. This area, called a mixing zone, is defined in the discharger's NPDES permit. For example, a typical mixing zone could be defined as that area within a one hundred-foot radius from the point of discharge. Water quality standards—except for acute toxicity—are suspended within the mixing zone, but they must be met at its boundary.

A water body is water quality limited when it does not meet a water quality standard or support recognized beneficial uses. Since this condition is not allowable under the Clean Water Act, states must take action to bring the water body back into compliance with the standard. As a rule, staff from the responsible state water quality agency performs studies to determine the cause of the violation. If the condition is caused by discharges into the water body, states limit these discharges through a process of establishing the total maximum daily load (TMDL), waste load allocations (WLAs), and load allocations (LAs). The TMDL is the total amount of a pollutant that can be discharged into the water body without causing an in-stream violation of the applicable water quality standard. The waste load allocations and load allocations are the quantities of pollutant that the respective point and nonpoint sources are allowed to discharge while meeting water quality standards. The TMDL consists of these allocations plus an amount set aside for reserve.

The public has the opportunity to become involved in the process of controlling discharges to water quality limited waters.

Agency staff first develops a draft plan outlining how discharges will be reduced to achieve compliance with the standard. They then circulate this plan to the public and to special interest groups for review and comment. Agencies usually schedule a public hearing to give interested people the opportunity to comment on the proposed plan.

You can also get involved in the periodic review of state water quality standards. According to the Clean Water Act, water quality standards must be reviewed every three years and the review process must be open to the public. You can request the draft standards from your state's environmental agency, review them, and express your views at scheduled public meetings or submit your comments in writing. You can call your state water quality agency or check their web site to learn about opportunities for public review and comment on water quality standards and TMDL studies.

Water Quality Certification

State water quality agencies must certify that activities subject to federal agency permits or licenses are in compliance with state water quality regulations. Section 401 of the Clean Water Act requires this procedure, commonly referred to as the 401 Certification process. For example, a dam operator must secure a license from the Federal Energy Regulatory Commission (FERC) to construct and operate a power generating dam. The state water quality agency must review this proposed activity and issue a 401 Certification—also called Water Quality Certification—before FERC can issue a license to the dam operator.

To request certification, an applicant prepares a plan that describes the project and its potential effects on water quality and submits it to the state agency. Based on the information in the plan, and compliance with state water quality standards and regulations, the state can approve or deny the request.

Wetlands Protection

As mentioned in Chapter 2, wetlands are disappearing rapidly throughout the world. Fortunately, we have slowed the rate of loss in the United States by enacting wetland protection regulations at all levels of government.

One of the most commonly applied regulatory programs, administered by the United States Army Corps of Engineers, is the wetland permitting program. In 1972, Congress codified these regulations in Section 404 of the Clean Water Act. Before anyone can legally discharge dredged or fill materials into waters of the United States—which includes most wetlands—they must obtain a Section 404 permit from the Corps. The Corps issues these permits based on the merits of the application and compliance with Section 404 regulations.

Again, because this is a federal permitting program, the state where the activity is planned must also certify that it will not cause violations of state water quality standards, as required by Section 401 of the Clean Water Act.

In many states, land boards and similar agencies are also involved in regulating wetlands. For example, the Oregon Division of State Lands works with the Corps to implement both federal and state

Fresh water wetland, Umpqua River, Oregon

wetland protection regulations. If a person proposes to remove, fill, or alter more than fifty cubic yards of material within state waters, including wetlands, they must obtain a permit from the Division of State Lands and/or the Corps. These two agencies have worked together to develop a single permit application form that the applicant completes and submits to both entities.

One of the practical difficulties in implementing wetland regulations is determining what constitutes a wetland. Is a wetland always a wet, swampy area? The short answer is no. Wetlands come in many varieties and go by different common and scientific names: bogs, swamps, marshes, emergent wetlands, forested wetlands, wet meadows, saltwater marshes, wooded swamps, scrub-shrub wetlands, riparian or shoreline wetlands, and so forth. According to the federal regulations implementing Section 404 of the Clean Water Act, wetlands are:

> *Those areas that are inundated or saturated by surface or*
> *ground water at a frequency and duration sufficient to support,*
> *and that under normal circumstances do support, a prevalence*
> *of vegetation typically adapted for life in saturated soil*
> *conditions.*

Wetland scientists commonly use three criteria to identify and determine the boundaries of wetlands: 1) hydrology, which is the absence or presence of water, 2) the occurrence of wetland plant species, called hydrophytes, and 3) the existence of hydric soils. The Corps has developed a special federal manual to assist scientists in correctly identifying wetlands.

In 1977, President Carter issued Executive Order 1190 to prevent wetlands from being harmed by federal activities. This order directs federal agencies to avoid activities that harm wetlands unless there are no practicable alternatives. The order states in part:

> *Each agency, to the extent permitted by law, shall avoid*
> *undertaking or providing assistance for new construction located*
> *in wetlands unless the head of the agency finds that 1) there is*
> *not a practicable alternative to such construction and 2) the*

proposed action includes all practicable measures to minimize harm to wetlands which may result from such use.

In 1986, Congress enacted the Emergency Wetlands Resource Act. This act requires the Secretary of the Interior, acting through staff of the United States Fish and Wildlife Service, to evaluate and report on the status and trends of wetlands in the United States every ten years or so. These ten-year reports give us a broad look at our wetlands and they have not painted an encouraging picture. We are continuing to lose wetlands and their valuable water quality benefits in spite of state and federal wetland protection regulations.

State Environmental Policy Act

Other environmental policies, such as the State Environmental Policy Act (SEPA), do not focus on protecting individual components of the environment like wetlands. Rather, their aim is to protect the entire natural and human environment.

Some states have a process for reviewing the environmental impacts of actions taken at the state and local level. This process generally follows the same basic approach as NEPA. In fact, sometimes these state regulations are affectionately referred to as "baby NEPA" requirements.

Most SEPAs require a project sponsor to describe the proposed action and its alternatives, and summarize expected environmental effects. Some jurisdictions have streamlined this review process by developing an environmental checklist or questionnaire. Regardless of their format, however, these state environmental reviews always include an evaluation of the proposed project's effect on water quality.

Local Regulations

In addition to state and federal programs, local programs have been developed to protect water quality in our communities. Sometimes these local programs seem especially pertinent because they focus on problems closer to home.

Construction Related Ordinances

Cities and counties often adopt special ordinances to protect local waters from construction-related problems. For example, many of them have adopted ordinances to keep soils and construction materials from entering waters near construction sites. These ordinances require erosion control features such as sediment fences, inlet protection, gravel entrance pads, and revegetation to be designed into a project and shown on the construction drawings. They also require construction contractors to maintain these erosion control features and manage their building materials properly. Contractors must keep the erosion control measures in good working order, since many jurisdictions inspect construction sites for compliance. Local agencies can revoke building permits if the builder does not comply with these erosion control and materials management ordinances.

Sometimes water quality protection measures are associated with land use approvals. In areas with strict land use laws, for instance, the local land use authority must review and approve proposed projects before developers can move forward. The project sponsor may be required to identify potential water quality impacts as part of the land use approval process. Sponsors usually must develop plans to eliminate or minimize these impacts before the project is approved.

Most local jurisdictions have flood plain ordinances intended to prevent flooding that also benefit water quality. These ordinances are often modeled after guidelines established by the Federal Emergency Management Agency (FEMA). By following FEMA guidelines, communities are allowed to participate in the National Flood Insurance Program, making them eligible to receive federal assistance if a flood occurs. These ordinances also serve to protect water quality by preventing flood plains from being randomly developed and disturbed. The vegetation and soils in the flood plains, and adjacent wetlands, provide important water quality benefits. They remove pollutants from waters passing through them. Vegetation also benefits water quality by shading the water, keeping it cool.

Some local jurisdictions require proposed construction projects to be reviewed during the design phase of the project. During this design-phase review, city or county staff may require design modifications to prevent the project from harming water quality.

Special District Ordinances

Communities sometimes form special service districts to address specific water quality concerns. Unlike cities and counties, whose responsibilities include everything from roads and parks to schools and fire protection, these special districts have limited obligations. Their efforts often can be focused at managing water quality in a single body of water, or in a body of water and its tributaries.

For example, in Lincoln City, Oregon, the Devils Lake Water Improvement District was formed to help address the specific water quality problems of Devils Lake. Because the lake suffers from an abundance of aquatic weeds, the district's staff developed a novel, site-specific program to control weeds. They introduced a non-reproductive species of weed-eating fish into the lake. They also routinely monitor the water chemistry of the lake to evaluate the effect of their lake restoration activities.

A special governmental body in the Seattle, Washington area, called the Puget Sound Water Quality Authority, helps maintain and restore the quality of the water and other natural resources in Puget Sound and its tributaries. This group works with cities, counties, state agencies, and the public to accomplish its mission.

These special districts provide technical and managerial guidance about water quality to the public and local jurisdictions. They may publish technical guidance manuals and model ordinances. In some cases, they may have the authority to develop and implement their own rules and regulations such as ordinances for controlling erosion, stormwater runoff, and other sources of water pollution. One of these special service districts may exist in your area. If so, you may want to give them a call. They will likely provide you with useful information about water quality issues and concerns in your community.

Summary

This chapter introduced you to important water quality rules and regulations like the Clean Water Act and the Safe Drinking Water Act. Refer to the references in Additional Reading   below for additional information. The most detailed reference information on federal programs is provided in a set of documents called the Code of Federal Regulations (CFRs). The CFRs, which contain the letter of the law, are routinely updated to reflect amendments and other changes. You can find the CFRs in the governmental document section of most public libraries or on one of the many web sites listed in Chapter 8 of this book.

Even with our many federal, state, and local rules and regulations, some of our attempts to restore and maintain the quality of our waters have fallen short of the mark. Some of the waters in the United States are still polluted in spite of our continued efforts to control individual sources of pollution. The next chapter introduces you to watershed management, a way of thinking more broadly about water quality protection.

Additional Reading

Adler, R. W., Landman, J. C., and D. M. Cameron, 1993. *The Clean Water Act 20 Years Later*. Natural Resources Defense Council, Island Press, Washington, D.C.

Chadwick, D. H., 1995. *Dead or Alive, The Endangered Species Act*. National Geographic, Vol. 187, No. 3, March 1995.

Cheremisinoff, P. N., and A. C. Morresi, 1979. *Environmental Assessment and Impact Statement Handbook*. Ann Arbor Science Publishers, Inc., Ann Arbor, Michigan.

Code of Federal Regulations, Title 40, Parts 100 to 149 (40 CFR 100-149), Protection of the Environment.

Dahl, T. E. and C. E. Johnson, 1991. *Status and Trends of Wetlands in the Conterminous United States, Mid-1970's to Mid-1980's*. United States Dept. of the Interior, Fish and Wildlife Service, Washington, D.C.

Environmental Protection Agency, 1976. *Quality Criteria for Water* (the "Red Book"). Environmental Protection Agency, Washington, D.C.

Environmental Protection Agency, 1986. *Quality Criteria for Water* (the "Gold Book"). Environmental Protection Agency, Washington, D.C. (EPA 440/5-86-001).

Environmental Protection Agency, Region VIII, 1993. *Everything You Wanted to Know about Environmental Regulations but Were Afraid to Ask: A Guide for Small Communities, Revised Edition.*

National Research Council, 1995. *Science and the Endangered Species Act.* National Academy Press, Washington, D.C.

U.S. Congress, Office of Technology Assessment, 1984. *Wetlands: Their Use and Regulation.* U.S. Government Printing Office, Washington, D.C.

Water Pollution Control Federation, 1982. *The Clean Water Act with Amendments.* Published by the Water Pollution Control Federation (now called the Water Environment Federation), Washington, D.C.

6. The Watershed Approach

Salmon River, Idaho

Over the past thirty years, the United States has made a great deal of progress in reducing water pollution from municipal and industrial point source discharges. We are also starting to make progress in controlling more diffuse, nonpoint sources of water pollution, such as stormwater runoff. Yet, even with our efforts and accomplishments, many of the nation's waters have not met the fishable and swimmable goals of the Clean Water Act. Why are some of our waters still polluted? Why were some of our earlier efforts unsuccessful? How do we approach and solve these problems now?

Many of our water pollution control efforts have been scattered and disconnected because we have focused on individual scenes rather than the big picture. If we were doctors, we might be accused of concentrating on fixing individual ailments instead of attending to the total health of our patient. A new, broader approach to water pollution control is emerging to overcome some of the shortcomings of our earlier efforts. This new approach focuses our water quality and water pollution control efforts in watersheds.

Fisher Creek Watershed, Washington

What is a watershed? It is essentially a drainage basin. You can think of it like a kitchen bowl or basin, but on a much larger scale. If you fill a kitchen basin with water and then make a hole in its bottom, all of the water in the basin will run "downhill" and drain out of the hole. Similarly, a drainage basin or watershed is a large basin of land. When rain falls into the basin, all the runoff will run downhill and drain to the river, stream, or lake at the bottom of the basin. For example, the Columbia River watershed is the portion of land that drains into the Columbia River. It stretches from British Columbia to the Pacific Ocean and includes parts of Washington, Oregon, Idaho, and Montana. All precipitation falling into this basin eventually enters the Columbia River through surface or groundwater flow and ultimately drains into the Pacific

Ocean. Moreover, all activities occurring within this watershed can affect water quality in the Columbia River.

Individual tributaries of larger rivers have their own watersheds. For instance, in Wisconsin, the Namekagon River watershed consists of all the lands draining into it. Since the Namekagon is a tributary of the St. Croix River, its watershed is also part of the larger St. Croix River watershed. Both of these rivers are tributaries of the Mississippi River and their watersheds are part of the huge Mississippi River watershed. We call these smaller watersheds within larger watersheds subwatersheds or subbasins.

Watersheds begin at mountaintops or ridges and they extend to the points of lowest elevation in the basin. Precipitation falling in a watershed flows downhill from the mountaintops and ridges— the upper boundaries of the watershed—to the streams and lowlands below. Watersheds are made up of all the landforms, vegetation, organisms, and water bodies in the basin. They may consist of uplands, lowlands, areas above and below timberline, forested areas, agricultural areas, urban areas, streams, wetlands, lakes, estuaries, and all the organisms living in these areas.

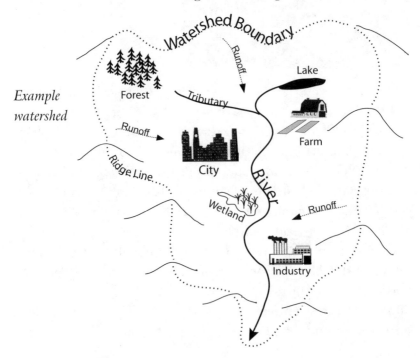

A watershed is a well-defined area for directing water pollution control efforts because all activities occurring within it have the potential for affecting water quality in that basin. Activities such as logging, farming, road building, mining, home construction, and others can affect water quality. Natural events occurring in a watershed, such as forest fires, land slides, flooding, and erosion also influence water quality. Activities occurring at any given point in the basin have the potential for reducing the quality of the waters downstream.

In the past, we focused many of our water pollution control efforts on individual sources of pollution, such as municipal or industrial discharges. With the new watershed approach, we evaluate all sources of pollution in a watershed collectively. That way, the benefits gained by strict control of point source discharges are not undone by pollution from other sources elsewhere in the watershed. This approach allows us to direct our water pollution control efforts within the watershed in ways that achieve the greatest benefit for the least cost. Our efforts can focus on achieving total watershed health, not just health in one part of the system. Moreover, we can evaluate the results of individual and combined efforts in light of the overall goal of achieving a healthy watershed system.

One of the many positive outcomes of evaluating water quality from a watershed perspective is the increased opportunity for public involvement. Local residents can assist engineers, scientists, and other environmental professionals in studying the watershed and fixing problem areas. Residents can get involved in everything from collecting water quality samples to planting trees along streams to provide shade and prevent erosion. By being involved, they develop a sense of stewardship for their own watersheds.

Do you know what watershed you live in? You can answer this question by looking at a map and finding the nearest creek that would receive runoff from your property. Then, follow this creek to where it enters the next largest stream or river and so forth. These creeks and rivers and the basins of land that drain into them are your watershed.

The remainder of this chapter summarizes the typical steps involved in using the watershed approach for protecting water quality. In its simplest terms, this approach consists of characterizing the watershed, identifying activities that influence water quality in the basin, developing alternatives to address problems, selecting and implementing the best alternatives, and monitoring the watershed to determine the results of water quality improvement efforts.

Johnson Creek
Watershed, Oregon

Watershed Characteristics

To understand the overall water quality picture, investigators studying the watershed must first determine some of its physical, chemical, and biological characteristics. They define many of the physical characteristics of a watershed by answering the following questions: How large is the watershed? What are its boundaries? What type of terrain exists in the watershed? What is the geology of the area? Where are the streams located? Do lakes or ponds exist in the watershed? Do wetlands exist? What temperature are the waters? Are the banks of the streams in good condition, or are they eroding?

A review of a watershed's physical characteristics may also include evaluating information about the climate: such as precipitation, evaporation, air temperature, relative humidity, and wind speed.

The chemical characteristics of the watershed include the chemical makeup of all important water bodies in the watershed—rivers, streams, lakes, wetlands, estuaries, and so forth. Investigators determine these characteristics by measuring water quality parameters, such as dissolved oxygen, pH, organics, solids, nutrients, and toxics. The mineral makeup of the soils also may be included as part of the chemical characterization of the watershed.

The biological characterization consists of an inventory of the plants and animals in the watershed, particularly aquatic plants and animals. The characterization identifies and evaluates sensitive species such as those that are threatened or endangered. For instance, the inventory may include a list of salmon spawning areas and an estimated count of the number of salmon returning to these spawning areas during the review period.

Today, many groups are using computerized data management systems called geographic information systems (GISs) to inventory the physical, chemical, and biological characteristics of watersheds. These systems allow users to store important data about the watershed in the computer and relate that information to its geographic position within the watershed. For example, after inventorying soils, researchers can enter soil types into the GIS according to their location in the watershed. The GIS allows them to generate a map of the watershed showing where the individual soil groups exist. Researchers also can review other important information about soil groups, such as their chemical structures and textures. They can do the same thing with water bodies and other features in the watershed. This method of managing data geographically helps those studying the watershed to visualize the connections between the different types of information, such as soils and water quality, or habitat and wildlife.

Activities Affecting Water Quality

The next step for investigators in evaluating the watershed is to identify the various activities affecting water quality. These activities are generally associated with land uses within the watershed and individual actions associated with such use. For example, land uses include forestry, agriculture, industry, and urban development. Individual actions might include include logging, application of pesticides and fertilizers, industrial discharges, municipal discharges, and construction projects such as road and home building. Chapter 3 includes descriptions of these and other possible sources of water pollution in more detail.

These land uses and individual actions are typically shown on a map of the watershed, such as those produced by a GIS. This type of map gives those studying the watershed a sense of space and geography. It helps to identify problem areas, develop restoration alternatives, and implement monitoring efforts.

Natural events like changes in climate, forest fires, flooding, and erosion also affect water quality. These natural events usually are beyond anyone's control, but those studying the watershed generally document their occurrence and take them into consideration in their overall evaluation of impacts to water quality.

While identifying the activities influencing water quality in the watershed, the investigators begin developing a list of people most likely to be affected by watershed management decisions. These people include area residents, municipal officials, industry representatives, environmental organizations, agency representatives, and others. We sometimes refer to this group as the stakeholders, because they have a stake or interest in the watershed. Ideally, these people participate in all phases of the watershed management effort.

Alternatives for Addressing Concerns

Once researchers have characterized the watershed and identified water quality concerns, they begin developing alternatives to address these problems, both individually and collectively. Chapter 4 includes descriptions of many of the alternatives available for addressing water quality concerns. For example, we can control the effects of municipal and industrial point source discharges by constructing good municipal and industrial wastewater treatment plants and operating them properly. We can minimize the effects of urban stormwater runoff by constructing stormwater treatment facilities as part of our drainage systems, and by implementing proper erosion control practices during construction.

Environmental professionals also have developed alternatives for addressing impacts from agriculture and forestry and other nonpoint sources of pollution. These alternatives generally consist of using best management practices (BMPs) to minimize water pollution from these sources. For instance, BMPs for agriculture may include minimizing the application of pesticides, herbicides, and fertilizers; using less harmful forms of these products; and using appropriate methods for crop rotation, harvesting, and tilling. BMPs for forestry may include minimizing road construction, leaving uncut buffer areas around stream corridors, and using selective logging instead of clear-cutting. Agencies like the United States Department of Agriculture are continuing to develop best management practices for minimizing harm to water quality from regulated activities such as agriculture and forestry.

Hedges Creek,
Oregon

Community planners may develop programs for reducing, reusing, and recycling materials to help improve water quality in the watershed. They may implement other nonstructural methods of control as well. For example, they may enact erosion control ordinances, special land use regulations, or flood plain protection ordinances if they do not already exist.

Before specific alternatives for protecting and restoring the watershed are selected and implemented, interested parties usually identify their goals. Water quality goals are easier to evaluate if they are specific, measurable, and well defined. For example, specific goals could be to: 1) reduce average summertime temperatures in the lowest one-mile reach of the watershed's main stream by two degrees, in ten years; 2) reduce turbidity by twenty percent during spring runoff in five specific tributaries, over the next five years; or 3) increase the number of resident cutthroat trout by ten percent in ten years throughout the entire watershed. A goal should state where, when, how much, and why. By defining goals clearly and making them measurable, investigators leave less room for interpretation or argument about what it means to achieve the goals.

Broad goals that cannot be defined so rigidly, such as increasing or maintaining the health and beauty of the watershed, are also important. However, one typically can rely on only general observations to see if these goals are being attained, and general observations may vary from one person to the next.

Once investigators identify watershed protection and restoration goals, and a range of alternatives for meeting these goals, they select and then begin implementing the best alternatives. To be successful, the alternatives should have the support of the people interested in the watershed: citizens, municipal officials, industry representatives, environmental organizations, agency representatives, and others. If these people participate in both analyzing the problems and creating the solutions, they are more likely to help fund, monitor, and otherwise support the restoration alternatives selected.

Monitoring the Watershed

Once alternatives for improving water quality are implemented, researchers begin evaluating how well their efforts are working. They monitor water quality in the watershed by collecting and analyzing water quality samples.

Lake Pend Oreille Watershed, Idaho

Before beginning, investigators outline their monitoring procedures in a watershed monitoring plan. Obviously, a carefully prepared plan will provide better results than random sampling. The procedures outlined in the plan typically include the purpose of the sampling; the time, location, and dates of the sampling; equipment needed; number of samples to be taken; and parameters to be analyzed. The monitoring plan includes a section on quality assurance and quality control that outlines acceptable methods for collecting, handling, storing, and analyzing samples. The plan also normally requires investigators to take duplicate samples and distilled water "blanks" to verify the quality of both the sampling and the analyses.

A monitoring plan reflects watershed protection and restoration goals because it is designed to evaluate success or failure in reaching them. For example, the study group may include biomonitoring as part of their plan. If one of the goals is to increase the population of fish in the watershed's streams, investigators must count fish to see if this goal is being met. Researchers are beginning to use various types of biomonitoring to evaluate the health of the aquatic environment. Sometimes they use the existence or abundance of specific organisms as biological indicators because some species are more or less tolerant to pollution than others. They can develop biological indexes of integrity, or health, based on these results.

When investigators find their goals are not being met, they should question themselves about each step in the process. Have they identified the watershed's characteristics correctly? Have they identified all activities that significantly affect water quality? Which alternatives for protecting and restoring water quality are working and which are not? The watershed approach is an iterative process. Investigators must continually work through the process to get the desired results. Moreover, positive change takes time. Results are rarely seen in a month or a year. Rather, it may take decades or longer to see measurable improvements.

Summary

This purpose of this chapter was to introduce you to the broad approach of controlling water pollution by looking at whole watersheds. You learned how investigators evaluate watershed characteristics and concerns and develop alternatives to address pollution. You also learned ways of monitoring the watershed to review improvement efforts. As you will see in the next chapter, one of the most important reasons for focusing on water quality in our watersheds is to protect our drinking water.

Additional Reading

Dopplett, B., M. Scurlock, C. Frissel, and J. Karr, 1993. *Entering the Watershed: A New Approach to Save America's Ecosystems.* The Pacific Rivers Council. Island Press, Washington, D.C.

Environmental Protection Agency, 1994. *A Watershed Assessment Primer.* Region 10 Watershed Section, Seattle, Washington, EPA 910/B-94-005.

Environmental Protection Agency, 1991. *The Watershed Protection Approach, An Overview.* Office of Water, EPA/503/9-92/002.

Environmental Protection Agency, 1987. *Biomonitoring to Achieve Control of Toxic Effluents.* Office of Water, EPA/625/8-87/013.

Flynn, K. C., and T. Williams, 1994. *Watershed Management Enters the Mainstream.* Water Environment and Technology, Vol. 6, No. 7.

James, A., and L. Evison (Eds.), 1979. *Biological Indicators of Water Quality*. John Wiley & Sons, Inc., New York, New York.

Lavigne, P., 1994. *Challenges in Watershed Activism*. River Voices, the Quarterly Publication of the River Network, Volume 5, Number 2.

MacDonald, L. H., Smart, A.W., and R. C. Wissmar, 1991. *Monitoring Guidelines to Evaluate Effects of Forest Activities on Streams in the Pacific Northwest and Alaska*. Environmental Protection Agency, Region 10, Seattle, Washington, EPA 910/9-91-001.

Powell, M., 1995. *Building a National Water Quality Monitoring Program*. Environmental Science and Technology, Vol. 29, No. 10.

Schueller, T. R., 1987. *Controlling Urban Runoff: A Practical Manual for Planning and Designing Urban BMPs*. Metropolitan Washington Council of Governments, Washington D.C.

The Wetlands Conservancy, 1995. *The Citizen's Regional Watershed Handbook*. The Wetlands Conservancy, Tualatin, Oregon.

Walesh, S. G., 1989. *Urban Surface Water Management*. John Wiley & Sons, Inc., New York, New York.

7. Drinking Water

Hardy Creek, Washington

Where do we get our drinking water? Most of us get it from our faucets at home or work, or we buy bottled water from the store. Many of us take this convenience for granted. We do not often stop to consider what happens to the water before it reaches our taps or the local market.

The water we use for drinking is no different than any of the other waters already described in this book. It is connected to all water on earth through the hydrologic cycle, making it subject to contamination in a number of ways. No independent pool of "drinking water" exists on the planet. Our drinking water comes from either groundwater or surface water sources in our watersheds. We normally must spend a great deal of time, effort, and money to make sure our drinking water is free from contamination and pleasing to our senses of sight, smell, and taste.

Drinking Water Sources

A community's drinking water supply comes from ground and surface water sources. Some communities use surface water as their primary source of drinking water and groundwater as a backup source. Others use either surface or groundwater exclusively.

Groundwater sources of drinking water come from water-bearing soil formations in our watersheds called aquifers. We store and retrieve water from an aquifer in much the same way we would from a bucket of sand. If you poured water into a bucket of sand it would trickle to the bottom of the sand, but not go through the bottom of the bucket. It would be stored in the lower portion of the sand. You could retrieve the water by placing a stiff perforated tube into the sand and providing suction on the top of the tube, similar to using a straw to drink soda from an ice-filled cup. Likewise, when precipitation falls on a soil formation resting on bedrock, water percolates through the soil and is stored in the lower reaches of the soil, above the bedrock. We retrieve water from an aquifer by digging or drilling a well, placing a perforated pipe into it, and then using a pump to retrieve the water from the bottom of the well.

We often obtain high quality drinking water from groundwater sources because these sources are protected from surface contamination by the soil mantle above them. Groundwater sources may have high concentrations of minerals, however, because they are constantly in contact with rocks and soil, and some of the minerals in the rocks and soil dissolve into the water. Because drinking water from groundwater sources is generally of

Seton Lake, British Columbia

high quality, water providers may need only to disinfect the water prior to delivery and use by the public.

Creeks, rivers, and lakes are also commonly used for drinking water. In rare cases, the ocean is used after salt is removed with desalination equipment. Drinking water is usually easier to collect from surface water sources than from groundwater sources. One might collect drinking water from a flowing river by simply placing a screened diversion pipe facing upstream into the river. Water flowing down the river would pass through the screen, enter the pipe, and continue flowing down the pipe to a collection pond or basin. Some communities collect drinking water from a lake by placing a screened intake pipe into the lake and then pumping the water from the lake to a collection basin.

Drinking Water Treatment

State and federal regulations require communities to treat and disinfect drinking water before distributing it to the public, because surface water sources are more vulnerable to contamination from activities occurring at the earth's surface. Water providers commonly use four simple processes to treat and disinfect drinking water before delivering it to the tap of the consumer.

First, they remove most of the large solids in the water, such as leaves, twigs, and sometimes fish, by passing the water through a coarse screen at the beginning of the treatment process. This screen is similar to a window screen or a fine mesh fence, but sturdier. The screen captures materials larger than the openings in the screen, which are normally about a quarter of an inch, or 6 mm, wide.

Second, they use a process called clarification to remove the finer particles of sand, silt, and organic debris that pass through the coarse screen. They place the water into a large, still basin called a clarifier and the fine particles settle from the solution due to gravity. Sometimes they mix chemical substances such as

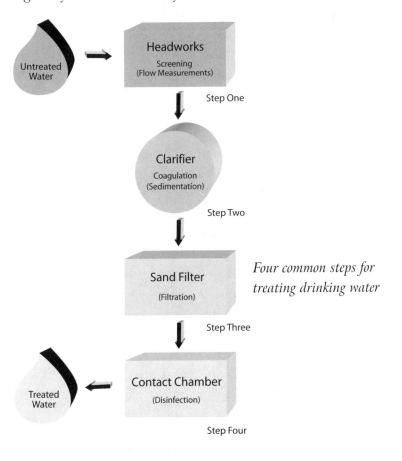

Four common steps for treating drinking water

coagulants and polymers with the water before it enters the clarifier to improve settling characteristics. These chemicals cause the individual fine particles to come together to form larger solids that settle more quickly.

Third, they pass the water through a filter—the most common type is made of sand—to remove the remaining fine particles that do not settle in the clarifier. They place the water on top of a large basin full of sand and allow it to percolate through the sand to the bottom of the basin. These basins are often rectangular. They may be ten feet wide by twenty feet long, or more, depending on the amount of water being treated. Water moves easily through the sand, but most of the solids are captured. The treatment plant operators periodically wash the sand filter to remove the collected solids by back-washing, which is a process where clean water is forced backwards through the sand. Sometimes, instead of back-washing, they scrape off and dispose of the top layer of the sand, revealing fresh sand below.

Finally, once the water has been screened, clarified, and filtered, the plant operators disinfect it to kill disease-causing, or pathogenic, organisms. Communities commonly disinfect drinking water by adding chlorine and mixing it thoroughly with the water—a process called chlorination. They may also disinfect the water by adding ozone, a process called ozone oxidation, or by passing the water through ultraviolet light, which is called UV irradiation.

Some communities provide additional treatment to remove specific dissolved substances not removed through the standard treatment processes. The most common substances of concern are calcium, magnesium, and dissolved organics.

Water providers may choose to remove calcium and magnesium because these minerals create hard water. Hard water leaves scaly deposits on plumbing fixtures and in industrial boilers. Hard water also makes it more difficult to clean clothes because soap does not work as effectively in hard water as it does in soft water.

Water softening is the process of removing calcium and magnesium. In the simplest type of softening, homeowners remove calcium and magnesium by passing water through a container of salt. As water passes through the container, the sodium in the salt trades places with the calcium and magnesium in the water, thereby removing these hard minerals and making the water softer. This chemical process is called ion exchange because sodium, calcium, and magnesium, which are all electrically charged substances called ions, exchange with each other when water comes in contact with the salt.

Because dissolved organic substances can cause color, odor, and taste problems, some communities also remove them. Removal is accomplished by passing the water through an activated carbon filter, which works like a sand filter except that water passes through a bed of activated carbon instead of a bed of sand. Activated carbon is similar to coal. It is a black, carbon material sold in granular or powdered form. Carbon is activated by heating it to a high temperature, which causes the material to fracture. Dissolved organics are sorbed by the many fractured surfaces as water passes through the filter.

Drinking Water Concerns

Many of us continue to be concerned about the quality of our drinking water, even with our ability to treat it to remove many contaminants. Water quality professionals call these concerns *source-related* if the source of our drinking water is threatened with contamination, *treatment-related* if the concern stems from the processes used to treat the water, or *distribution-related* if the distribution system carrying our water is in question.

Watershed Disturbances

Activities that disturb the land in our watersheds threaten the quality of our drinking water supply and are source-related concerns. Logging, road building, mining and other activities that involve digging into the soil or removing vegetation can be particularly harmful if they are not done carefully. Stormwater running over these disturbed areas can pick up sand, silt, clay, and other contaminants and carry them into our drinking water supply. Recall that both our surface water and groundwater supplies are fed by precipitation falling into our watersheds.

Drinking water intake,
E. Fork Santa Ana River,
California

Because of the threat of contamination, some communities have adopted rules or ordinances prohibiting land-disturbing activities in watersheds that supply their drinking water. Other communities allow some restricted activities in their watersheds, provided that these activities are done in a controlled and limited way. These communities customarily require proponents of any land-disturbing activity to implement proper erosion control techniques and spill prevention and control measures to prevent eroded soils or accidental spills from contaminating the water supply. Chapter 4 includes descriptions of specific measures used to control erosion and spills.

Why not just remove pollutants from our drinking water supply with proper treatment if watershed disturbances cause contamination? This approach would be shortsighted because treatment is expensive and not always effective. The cost of removing pollutants from our water supply is much higher than the cost of keeping them out in the first place.

Groundwater Contamination

Groundwater pollution can be associated with a variety of substances, such as industrial chemicals, fertilizers and pesticides, or animal and human wastes. These materials can percolate into the soil and contaminate the groundwater below if not managed properly. Once in the groundwater, they may be difficult or impossible to remove.

The threat of groundwater contamination is currently a concern in several cities across the United States. For example, Portland, Oregon's main water supply comes from surface water in the protected Bull Run Watershed located on the slopes of nearby Mount Hood. However, the backup water supply that provides supplemental water during drought conditions comes from groundwater wells lower in the valley. Recently, investigators discovered that groundwater near these wells was contaminated with the organic solvent trichloroethylene (TCE). An expensive cleanup effort is now underway to remove the contamination and protect the groundwater supply. Fortunately, these groundwater wells are not Portland's primary source of drinking water.

In other parts of the United States, groundwater is contaminated with inorganic substances, such as nitrates, that are found in sewage, animal wastes, nitrogen fertilizers, and food processing wastes. Nitrate contamination is a problem in Iowa, Kansas, Minnesota, Nebraska, and South Dakota. The Environmental Protection Agency estimates that, in these states, one out of every four private wells has excessive levels of nitrate.

High concentrations of nitrate cause two primary health concerns. First, as discussed in Chapter 2, infants given water or formula containing high concentrations of nitrate may develop a disease called methemoglobinemia, which can be fatal. Second, high concentrations of nitrate pose a potential cancer risk to the general population. The body converts ingested nitrates into carcinogenic compounds called nitrosamines. Researchers have found that nitrosamines cause cancer in laboratory animals. They are still investigating the relationship between high concentrations of nitrate and cancer in humans.

Microbial Contamination

The contamination of drinking water by microorganisms is a serious health concern worldwide. In fact, only about twenty-five percent of the world's population has safe drinking water. Doctors attribute many illnesses and deaths in developing nations to improper sanitation and contamination of the water by the microorganisms found in human feces.

Bacteria, viruses, and protozoans are all microorganisms capable of contaminating our drinking water and causing disease. For instance, bacteria are responsible for typhoid, paratyphoid, salmonellosis, shegellosis, bacillary dysentery, and Asiatic cholera. Viruses are responsible for infectious hepatitis and poliomyelitis. Protozoans are responsible for amebic dysentery and Giardiasis.

Typhoid fever is caused by the bacterium *Salmonella typhi*. Although typhoid fever is no longer a common disease in the United States, it was one of the major causes of death decades ago, and it continues to be a threat in developing nations. Typhoid fever causes a high fever, diarrhea, and ulceration of the small intestine. It is highly contagious. The expression "typhoid Mary" came as a result of the transmission of typhoid bacteria by a cook who moved across the United States. She was a carrier of typhoid, and transmitted the disease to her customers by contaminating their food and water.

The hepatitis A virus causes infectious hepatitis. Symptoms of this disease include fever, weakness, nausea, vomiting, abdominal cramps, and enlargement of the liver. The hepatitis virus can be present in drinking water contaminated with feces. Thousands of cases of infectious hepatitis are reported every year in the United States.

Public health professionals are currently investigating Cryptosporidiosis, a diarrheal disease caused by the parasite *Cryptosporidium parvum*. They are also studying Giardiasis, the disease caused by the Giardia protozoan *Giardia lamblia*. Although this disease is not generally life threatening, it does result in very uncomfortable, flu-like symptoms and diarrhea that may last for ten days or more. The Giardia protozoans form cysts that are resistant to disinfection. The relatively large size of Giardia, however, makes them easy to remove with proper filtration. Giardiasis is a common malady for hikers and campers who drink untreated and unfiltered water. Wild animals like beaver and deer carry Giardia in their feces, and water that comes into contact with their waste becomes contaminated.

We cannot test our drinking water for the presence of every known disease-causing organism. It would be impractical; testing would take too long and it would be too expensive. Instead, we test for the presence of these organisms by the use of indicator species. The most common indicator species are the fecal coliform group of bacteria. This group includes organisms like the bacterium *Escherichia coli*, which is always present in feces. Since coliform bacteria occur in large numbers in waters contaminated with feces, their presence in a water sample indicates that the sample may be contaminated with feces. Their presence does not prove that the water is contaminated but it does give cause for concern, especially if they occur in large numbers.

Clearly, we must protect our drinking water from microbial contamination to prevent the outbreak of disease. We can provide protection by keeping our water free from contamination and by disinfecting it to kill the harmful organisms prior to delivery.

Chlorinated Organics

The methods we use to treat our drinking water sometimes cause the formation of potentially harmful substances and become treatment-related concerns. For example, to kill microorganisms, we commonly disinfect our drinking water with chlorine using the chlorination process described in Chapter 4.

Unfortunately, chlorination can sometimes cause the formation of harmful chlorinated organic substances called trihalomethanes (THMs). Trihalomethanes are generated when chlorine combines with organic molecules found in many water supplies. These organic molecules come from the breakdown of vegetation and other organic materials found in the water. THMs cause cancer in laboratory animals, an indication that they may cause cancer in

"Bottled water" trucks

humans. Researchers are continuing to study the effects of THMs on human health.

Because of the problems with chlorine, communities are considering other methods of disinfection, such as passing the water under ultraviolet light or mixing the water with ozone.

Copper and Lead

Copper and lead pose a serious distribution-related problem. Water moving through distribution pipes and home plumbing on the way to the consumer may dissolve and pick up copper and lead along the way. The sources of copper and lead are the pipes, which until recently were manufactured using copper and lead; pipe linings; joints; and the solder used to hold the pipes together.

Long-term exposure to lead, even in low concentrations, can cause improper brain functioning and development. Scientific studies have clearly established the link between lead intake and the intellectual impairment of children. Unlike some contaminants, lead does not flush out of the body with body fluids. Once ingested, it remains in the fat and body tissue for life.

Copper is generally less of a concern than lead, but some studies indicate that prolonged ingestion of high concentrations of copper may result in liver damage. We do not yet clearly understand the effects of prolonged ingestion of low concentrations of copper, however. Regardless of its toxic effects, copper is often undesirable because it can impart a bad taste to drinking water.

Because of the concerns about copper and lead in drinking water, the United States Environmental Protection Agency has prepared information to help the public better understand and evaluate potential problems. Some of this information is available through their web site at the address listed in Chapter 8.

You can do three simple things at home to reduce your potential exposure to lead in your drinking water. First, let your water run for a minute or two in the morning when you first use it, or if it hasn't been used for several hours. Stagnant water left standing in your pipes has a greater chance to dissolve and pick up any lead the may exist in your home plumbing. By flushing out this stagnant water, you replace it with fresh water containing less lead. Second, use water from your cold water tap for drinking and cooking. Hot water is more corrosive; it can dissolve lead in your home plumbing more easily than cold water, causing a higher concentration of lead in your drinking water. Third, never use materials containing lead for making home plumbing repairs. Do not use lead solder, pipes, or connections.

Drinking Water Standards

What concentrations of specific contaminants are safe for humans? As mentioned in Chapter 5, the Safe Drinking Water Act (SDWA) helps answer this important question by establishing the MCLs allowable in a drinking water supply. The maximum contaminant

Water storage tank

levels that protect human health are called the Primary Drinking Water Standards and they have associated goals and action levels.

The Primary Drinking Water Standards for selected inorganic, organic, and microbial contaminants appear below. Recall that the concentration term mg/L is equivalent to parts per million, as described in Chapter 2. Milliliter is the term used to describe one one-thousandth of a liter and is abbreviated by the term ml.

Inorganic Chemicals (all in mg/L):

Antimony	0.006
Barium	2.0
Cadmium	0.005
Chromium	0.1
Copper	1.3
Cyanide	0.2
Fluoride	4.0
Lead	0.015
Mercury	0.002
Nickel	0.1
Nitrate (as N)	10
Selenium	0.05

Organic Chemicals (all in mg/L):

Alachlor	0.002
Benzene	0.005
Lindane	0.0002
Methoxychlor	0.04
Polychlorinated biphenyls (PCBs)	0.0005
Toxaphene	0.003
Trichloroethylene (TCE)	0.005
2,4-D	0.07
2,4,5-TP	0.05
2,3,7,8 TCDD (dioxin)	3×10^{-8}

Microbial Contaminants:

Giardia lamblia	0 (goal)
Viruses	0 (goal)
Legionella	0 (goal)
Total coliforms	0 (goal)

The Safe Drinking Water Act sets the goal for microbial contaminants at zero, since there are no known thresholds for these contaminants. Under certain circumstances, a single virus ingested with drinking water might cause disease. For practicality, however, the act requires water providers to remove approximately 99.9 percent of most microbial contaminants.

The SDWA also establishes Secondary Drinking Water Standards to protect drinking water from contaminants that primarily affect the aesthetic qualities of water, such as color, taste, and odor. Some of the Secondary Drinking Water Standards are listed below.

Contaminant:
Chloride	250 mg/L
Color	15 color units
Copper	1.0 mg/L
Corrosivity	noncorrosive
Fluoride	2.0 mg/L
Iron	0.3 mg/L
Manganese	0.05 mg/L
Odor	3 threshold odor number
pH	6.5 - 8.5
Sulfate	250 mg/L
Total dissolved solids	500 mg/L
Zinc	5 mg/L

The Primary and Secondary standards listed above provide a summary of the MCLs for some of the most important contaminants. However, other standards and special conditions also apply. EPA's web site for drinking water (www.epa.gov/safewater/mcl/html) provides a more detailed listing and explanation of these standards.

If you are concerned about the quality of your drinking water, you can have it tested. Some communities provide this service free. You also can hire a private laboratory to test your water. Businesses offering this service appear in the telephone directory under laboratories or water testing services. Three simple and relatively inexpensive tests that will provide you with useful information are the tests for bacteria, nitrate, and lead. After getting your results, compare them with the drinking water standards.

Water storage tower

Summary

This chapter introduced you to ground and surface water sources of drinking water and to some of the activities that threaten their quality. It also introduced you to methods of treating water to make it suitable for drinking. As you continue to learn about clean water, you may choose to get more personally involved in protecting it. The final chapter of this book tells you how.

Additional Reading

Basatch, R., 1998. *Waters of Oregon: A Source Book on Oregon's Water and Water Management*. Oregon State University Press, Corvallis, Oregon.

Bureau of Reclamation, 1985. *Ground Water Manual*. United States Government Printing Office, Washington, D.C.

Clark, J. W., Viessman, W., and M. J. Hammer, 1977. *Water Supply and Pollution Control, Third Edition*. Harper & Row Publishers, Inc., New York, New York.

Faust, S. D., and O. M. Aly, 1983. *Chemistry of Water Treatment*. Butterworth Publishers, Woburn, Massachusetts.

Fitts, C.R., 2002. *Groundwater Science*. Academic Press, San Diego, California.

Gray, N. F., 1994. *Drinking Water Quality: Problems and Solutions.* John Wiley and Sons, Inc., New York, New York.

Maier, R.M., Pepper, I.L, and C.P. Gerba, 2000. *Environmental Microbiology.* Academic Press, San Diego, California.

Sanks, R. L., 1980. *Water Treatment Plant Design for the Practicing Engineer.* Ann Arbor Science Publishers, Inc., Ann Arbor, Michigan.

Smethurst, G., 1988. *Basic Water Treatment for Application Worldwide.* Thomas Telford Ltd., London, England.

Steel, E. W., and T. J. McGhee, 1979. *Water Supply and Sewerage, Fifth Edition.* McGraw-Hill Book Company, New York, New York.

Tchobanoglous, G. and E. D. Schroeder, 1985. *Water Quality: Characteristics, Modeling, Modification.* Addison-Wesley Publishing Company, Reading, Massachusetts.

Viessman, W., and M.J. Hammer, 1985. *Water Supply and Pollution Control.* Harper Collins Publishers, New York, New York.

8. Getting Personal about Clean Water

Chilliwack River, British Columbia

What can an individual do to help keep our waters clean? You have already taken the first step by reading this book and learning the basics of water quality and water pollution control. Next, you can put this knowledge to work by getting involved in water quality protection at home and in your community.

The previous chapters of this book introduced many of the scientific and legal aspects of water pollution control. This chapter offers some ideas about how to begin applying what you have learned.

Water Quality at Home

The easiest and most personal place to get involved in water quality protection is at home, since many of your everyday decisions affect the quality of the water in your community. This section describes ten simple ways you can contribute to cleaner water.

Use Environmentally Friendly Cleaning Products

Biodegradable and nontoxic household cleaning products are better for water quality than poisonous cleaning products. They are also better for your family's health. Many environmentally friendly cleaners are now on the market. Read labels when buying cleaners and look for those that are biodegradable and nontoxic. Beware of products with skull and cross bones or similar warnings that tell you the contents are poisonous. Poisonous cleaners are a hazard to both human health and the environment.

Use any toxic cleaning products you may have appropriately and completely so you do not need to dispose of them. If you have leftover cleaners, dispose of them properly. Do not flush them down the toilet or throw them in the trash.

Flushing toxic cleaning products down the toilet is inappropriate because municipal treatment plants are designed to treat sewage, not toxic waste. Toxic cleaners can cause upsets at a treatment plant, reducing its ability to provide proper treatment. Some toxic products may pass through the plant without being treated at all. These pass-through pollutants end up being discharged to the river, where they cause water pollution, or contaminating the sewage solids. If your community applies its sewage solids to the land as fertilizer, these pass-through pollutants can adversely affect the soil and groundwater.

Rather than putting toxic products out with your regular trash, check with your disposal company to find out what provisions have been made for collecting household hazardous wastes. You

Ice fishing on Pineview Reservoir, Utah

can usually take these wastes, such as toxic cleaning chemicals and paints, to special collection sites. Community employees then take responsibility for properly managing and disposing of these waste products. Some can be recycled, while others must be disposed at government-regulated waste disposal sites.

Use Household Water Wisely

You can help maintain water quality in your community by using water wisely at home. Your use of water for drinking, washing dishes and clothes, taking showers, flushing toilets, and caring for your lawn and garden all have an effect on water quality.

Water providers invest a great deal of time, effort, and money to make sure we have a safe supply of water for drinking and other household uses. They filter and disinfect the water before delivering it to individual households through pipes and pumps. They hire professionals to design, construct, and operate the treatment facilities to ensure the water is safe. They construct laboratories and run analytical tests on the water supply to ensure that contaminants are not present. Our communities spend millions of dollars to provide us with a safe supply of drinking water.

If you waste drinking water, you are indirectly threatening water quality. When residents of a community use more water than necessary or planned, the community must either spend more money or provide poorer quality water. For instance, assume a community has the staff and treatment facilities to provide its citizens with one million gallons of safe water for drinking and other household uses each day. If the demand became two million gallons per day because everyone used twice the amount they really needed, the community would either have to spend more money for additional treatment or provide lower quality water to its citizens.

For most of us, all the water we use for washing our dishes and clothes, taking showers, and flushing our toilets goes from our house to the local sewage treatment plant. If treatment plants are overloaded, they do not operate efficiently, and the treated water that flows back into the river is of poorer quality. If we send more wastewater to the treatment plant, operators are forced to use more chlorine to disinfect the larger volume of water, which results in additional discharges of harmful chlorine into the environment.

The water you use in lawn and garden maintenance may pick up pollutants such as fertilizers, pesticides, and soil particles. Runoff from excessive irrigation can wash these pollutants off your property, into the stormwater system, and ultimately into our rivers and streams.

A finite amount of clean water is available in the environment. By taking clean water out of our rivers and streams to supply our water needs at home, we reduce the amount of clean water available for other uses and other species. Moreover, the water we return to the ecosystem is almost always lower in quality than when we received it. This process of taking water of high quality and replacing it with water of poorer quality results in an overall reduction in the quality of water in the environment. The more wisely and efficiently we use our water at home, the cleaner our water will be for everyone and everything.

Use Household Energy Wisely

Hydropower and fossil fuels like coal, oil, and gas continue to supply most of our energy needs. All of these sources of energy have associated water quality concerns.

Researchers have linked the declines in water quality and habitat on major rivers, such as the Columbia River in the northwestern United States, to dams used for hydropower. These reductions in water quality and habitat are threatening some populations of fish. Using energy more efficiently at home will require fewer dams to provide hydropower to meet our energy needs. Minimizing our need for energy will allow us to operate essential dams in ways that maximize resource protection, not just energy production.

Bonneville Dam, Columbia River, Oregon

Building fewer dams, and managing existing ones better, will result in less water being diverted from our rivers, less evaporation, less habitat loss, lower water temperatures, and an overall improvement in water quality.

Using energy efficiently at home also will result in a reduced need for fossil fuels. Less need for fossil fuels will result in fewer oil spills, less offshore mining of oil and natural gas, fewer coal mines, fewer hydrocarbon emissions into the atmosphere, and an overall improvement in water quality.

Compost Your Lawn Clippings, Yard Debris, and Food Wastes

One of the biggest environmental challenges facing society today is the proper management of the huge amounts of solid waste we generate. Some large cities have so much solid waste they have literally created mountains of it. A ski hill near Detroit, Michigan, for instance, is referred to as Mount Trashmore because it was built on a mountain of trash.

Most communities dispose of the solid waste we leave at the curbside in landfills. These landfills are expensive to design, construct, and maintain. By reducing your generation of solid waste, you can help reduce the need for new landfills and save space in the existing ones. By generating less waste, we can operate our existing landfills in a more environmentally safe manner and use our valuable land resources for purposes other than waste disposal.

Even landfills that are carefully designed and operated pose some threat to ground and surface water quality. Water mixing with refuse in a landfill can pick up pollutants and create leachate (see Chapter 4). Moreover, if the landfill liner leaks, leachate can move down through the soil and contaminate groundwater. Communities that collect and treat their landfill leachate must ultimately discharge it into a stream or back onto the land, where it may run off and eventually reach surface water.

As mentioned earlier, the use of water by humans generally results in a reduction in its quality. For example, the rainwater that falls on a landfill is usually of very high quality. In uncovered landfills, this high quality rainwater filters down through materials in the landfill, picking up pollutants and creating landfill leachate. Even if this leachate is properly collected, treated, and disposed of, the resulting water will be poorer in quality than the original rainwater.

Composting your biodegradable household wastes is one way you can contribute to the goal of reducing the amount of solid waste that goes into our landfills. You do not need to dispose of your lawn clippings, leaves, and other yard debris or food wastes in the trash. These materials are all biodegradable and you can turn them into compost. Compost piles are easy to create and they require little maintenance. You can usually keep your household compost pile relatively small because materials in the compost are constantly becoming smaller through biodegradation. Since compost contains many nutrients, you can use it as a natural fertilizer for your landscaping needs. Your local library probably has several books on backyard composting available. Some of the web sites listed at the end of this chapter also have information about composting.

Recycle and Reuse Household Goods instead of Throwing Them in the Trash

Another way you can reduce solid waste is to recycle materials like newspaper, glass, tin cans, aluminum, magazines, office paper, waste paper, and some plastic products. The list of recyclable household materials continues to grow as we discover better recycling processes and more uses for recycled products. Many communities now sponsor recycling programs for their residents. Some of these programs require you to take your recyclable materials to a central location and others pick up the materials at your curbside.

In addition to reducing the amount of waste that we put into our landfills, recycling helps to conserve energy and natural resources. For example, aluminum manufacturing and paper production plants are extremely energy intensive and consume

Curbside recycling, Portland, Oregon

natural resources. Aluminum manufacturing plants are usually built near large rivers and run by hydropower. Using recycled aluminum reduces the need for hydropower and also reduces the environmental impact associated with removing raw aluminum from the earth. Similarly, recycling paper reduces the need to log timber from our watersheds, requires the use of fewer pulping and bleaching chemicals, and results in an overall benefit to water quality.

Reusing household goods also reduces the use of energy and natural resources and limits the amount of waste material placed in our landfills. You can help protect the environment by using items that can be washed and used again instead of used once and thrown away. You can use cloth napkins instead of paper towels and cloth diapers instead of disposable ones. You can find other uses for containers instead of throwing them away. For example, you can reuse peanut butter jars for storing bulk food items or milk jugs for storing emergency water supplies.

Follow Good Car Maintenance Practices.

A well-maintained car results in less water pollution. You do not have to go far on a rainy day before seeing the characteristic sheen of gas or oil spots on the roadway. Often, poorly maintained cars with leaking gas tanks or oil pans are responsible for these spots. These spilled petroleum products will wash off the surface of the road, into a nearby ditch or drainage pipe, and ultimately into a nearby waterway.

If you maintain your own car by changing the oil, antifreeze, or windshield wiper fluid, make sure you collect and dispose of the used fluids properly. Do not allow them to get into the water and cause pollution. If these liquids are poured down a storm drain, flushed down the toilet, or poured onto the soil, they can cause surface water or groundwater pollution. Many gas stations and quick-lube oil stations will accept your used automobile fluids because they can recycle them. The recycling businesses sometimes pay these stations for their used products.

Use Your Automobile Less and Use It More Selectively

Because using an automobile results in air pollution and water pollution, the more you use it, the more you contribute to the deterioration of water quality. Recall that air and water are linked by the hydrologic cycle, as described in Chapter 1. Contaminants in the air are picked up by water vapor in the atmosphere. These contaminants fall to the earth with precipitation and are transferred to the land and water. The phenomenon of acid rain provides an extreme example of the link between air pollution and water pollution.

The more we use our automobiles, the more need we have for gas and oil production and a host of related activities that have the potential to harm water quality, such as offshore oil exploration, mining, and transportation of oil and gas over our waterways. Instead of driving your car, use mass transportation. You can take the bus or the commuter train to work and support your community's mass transportation efforts. The bus and commuter train can provide the same transportation service as hundreds of

Crossing the Table River, British Columbia

cars. Or, you can join a carpool. One carload of commuters can take the place of four or more separate vehicles.

You can also use your automobile more selectively. Walk or ride your bike instead of hopping in your car to run a nearby errand. Reduce the amount of time you are on the road by completing several errands on one trip instead of making a separate trip for each. Consider doing your errands with friends and neighbors. Use phone or mail order to reduce your trips to the store or shopping mall. By making thoughtful transportation choices and using your automobile less and more selectively, you can contribute to protecting water quality and the environment, and also save money.

Be Mindful of Your Use of Pesticides, Herbicides, and Fertilizers in Home Landscaping

Like many people, you may use pesticides, herbicides, and fertilizers for your home landscaping needs. These products can help you have tall trees, lush lawns, and beautiful flowers. Unfortunately, they can also cause water pollution if they are over-applied or otherwise used improperly. When it rains, these products can be carried by stormwater runoff from your property into the nearby storm drainage system and the nearest stream. These products can also leach down through the soil and cause groundwater pollution.

If you use these products at all, use them carefully. Always follow the manufacturer's recommendations. Never over-apply pesticides, herbicides, or fertilizers. With these products, more is *not* better. Over-application results in wasted product that is then available to harm water quality, and possibly children and pets. To avoid having these products get into the storm drainage system, do not apply them in wet weather and do not apply them near storm drains or in other areas where they could easily run off your property.

Better yet, avoid or limit your use of these products by opting for more natural methods of landscaping and pest control. Instead of using pesticides, you can use birds, marigolds, ladybugs, nematodes, and mild soap to control pests. Instead of using herbicides, you can remove weeds by hand or use black landscaping fabric to cover areas and prevent weeds from growing in the first place. Instead of using packaged fertilizers, use homemade compost made from lawn clippings, food waste, and other biodegradable materials.

Educate and Involve Your Children, and Set a Good Example

Environmental education is more popular now than ever. Many of our children are learning about the environment through activities at school, scouts, and 4-H clubs. Our children are generally more interested and aware of the environment than we were at their age. We can reinforce their interest and awareness at home by reading to them about the environment. Read portions of this book to them. Get other books and videos about nature that are appropriate for their age.

Set an example for your children by applying the simple techniques described in this chapter: use water and energy efficiently, recycle, use pesticides and herbicides sparingly if at all, and use compost for fertilizer. By setting a good example for your children, you help educate them, reinforce what they learn at school, and contribute to protecting water quality and the environment.

You may want to involve your children in environmental protection at home as a way of making their education real. You can give them responsibility for helping to sort recyclables, let them help you with your household composting, and ask them to help conserve water. By allowing your children to get involved, you show them that you are interested in their environmental education.

Be Mindful of Runoff from Your Property

Precipitation falling on your property soaks into the soil, is taken up by your plants, or runs off into the storm drainage system or nearby ditch. To limit the pollutants in the stormwater running off of your property, you can take measures to reduce the pollutants on it. As discussed earlier, you can use fewer pesticides, herbicides, and fertilizers, and care for your automobiles properly so they do not leak oil and gasoline. You can also help reduce stormwater pollution by reducing the runoff leaving your property. For example, you can reduce the use of impervious materials like concrete by using gravel and sand-set bricks for walkways and patios. This practice allows rainwater to soak into the underlying soil. You can reduce runoff from rooftops by directing it into infiltration pipes placed in the yard, or by collecting the water in barrels and using it later for watering your lawn, plants, or garden.

These ten relatively simple ways of protecting water quality at home are only a beginning. You can probably think of many more, specific to your own home, if you stop to give it some thought. The most important thing to remember is that clean water begins at home. What *you* do at home makes a difference.

Public Involvement

You can also work for clean water by getting involved in water quality protection in your community. A recent television commercial uses a clever approach for selling perfume. In the commercial, an actress speaks in a low voice, saying, "If you want

Hawthorne Bridge, Willamette River, Oregon

to capture someone's attention, whisper." We strain to hear what she is saying. By speaking in a whisper, she draws our attention. An individual person's voice can be likened to a whisper. Sometimes it catches one's attention better than the roar of the crowd. An individual voice is personal. Rather than being the voice of an unidentifiable mass, it is the voice of a single, caring human being.

By speaking out individually, you tell the world, "I care enough about the quality of our waters to stand up and be counted; I am willing to take the responsibility to work towards protecting and restoring the water environment, which is so important to us all."

You can make a difference by writing letters to your local, state, and federal representatives to tell them about the specific water quality concerns that affect your community. They have the responsibility of answering your questions, keeping you informed, and listening to what you have to say. It is their job to listen so they can represent you properly. They also have staff to help them fulfill their responsibilities when they cannot carry out the tasks personally.

You can use this book for reference to add specific details that show you are informed. You may want to state your specific concerns and make reference to appropriate rules and regulations or the chemistry and microbiology involved. You can also describe your broader concerns about water quality and water pollution control.

Your local post office often posts the names and addresses of your local representatives. Some states print a "Blue Book" which includes listings of local, state, and federal representatives and how they can be contacted. You can also get this information at your local library or online.

Find the names of your United States Congressional representatives by dialing, toll free, the Federal Information Center at 1-800-688-9889, or at the online address listed in the next section of this book. Their local addresses and phone numbers are also included in the governmental listings in the phone book. The addresses of the United States President and Congress in Washington are listed below:

The White House, 1600 Pennsylvania Avenue, Washington, D.C. 20500

United States Senate, Washington, D.C. 20510

United States House of Representatives, Washington, D.C. 20515

Your voice will also be heard by voting according to your convictions. Inform yourself about water quality issues and find out where your elected officials stand. Listen and watch radio and television broadcasts where these issues are debated. A single phone call to your representative or their staff may be all it takes to find out how they represented you on a particular issue and why. You can use the power of your individual vote to show your support for good representation.

You can attend public meetings to voice your concerns or submit written comments during public review and comment periods. Many of the programs designed to protect water quality provide opportunities for this type of public involvement. For example, prior to issuing an NPDES permit, state agencies provide the opportunity for public comment on the draft permit. State agencies also review and revise water quality standards every three years. Once agency staff develops draft standards, they make these drafts available to the public for review and comment. Other programs like NEPA provide the opportunity for public review and comment on environmental assessments and environmental impact statements.

Often, the agency responsible for a particular environmental program issues notices to inform the public about specific actions being considered by the agency. For instance, public notices are prepared by most states before they issue NPDES permits. You can often get on an agency's mailing list to receive these types of notices by simply contacting the agency. Call and inform the receptionist of your interest and your call should be directed appropriately.

Environmental Organizations

Sometimes it takes more than one voice to get the clean water message heard. It often takes the voices of organized groups. Nearly all environmental organizations have an interest in protecting water quality, though it may not be their primary focus. You may want to consider joining one or more of these groups as another way of getting involved. Groups and organizations can help you stay informed about water quality and other environmental issues and they can be effective in representing members in the political and legal arenas.

Environmental groups play an important role in a democratic society. They represent the goals and beliefs of the founders of the group and its members. You should realize, however, that each of these groups has a different personality and agenda. Some groups lean to the conservative side and some are more liberal. Some are more willing to affect change by taking drastic measures. The key to finding a group that suits you is to find one with a personality like yours—a group with which you share common beliefs.

Canoeing on the Deschutes River, Oregon

One way of finding out about a particular group is to review a copy of the group's mission statement or statement of goals. What does the group believe in? Another way is to review a copy of the group's annual financial statement. What happens to the money members donate?

Your Resource Guide to Environmental Organizations by J. Seredich is a good resource for learning more about environmental groups (see **Additional Reading** below). This guide includes contact information, purpose, accomplishments, and membership benefits for many environmental organizations.

The listing of Internet resources below also includes contact information for environmental organizations. The author does not specifically endorse any of these groups, and this list is not complete. It is simply a place to get started. Many of these groups have local offices where you can get additional information.

Internet Resources

The Internet provides a wealth of information about water quality, water pollution control, and related topics. You can access the Internet through your home computer, using a variety of Internet service providers. Or, you may be able to access the Internet on computers at your local school or library. A teacher or librarian can probably assist you in accessing the Internet if you are not familiar with the process.

You should also be able to search for web sites based on your topic of interest. For example, you should be able to insert a topic such as "wetland treatment" into your computer's search engine and your computer will list web sites that are related to this topic.

Below are some useful Internet resources to get you started:

Drinking Water

American Water Works Association (awwa.org)
National Drinking Water Clearinghouse (estd.wvu.edu/nsfc)
Water Partners International (water.org)
Water Quality Association (wqa.org)
USEPA Office of Groundwater and Drinking Water
    (epa.gov/ogwdw)

Education

Americas Clean Water Foundation (acwf.org)
Global Rivers Environmental Education Network
    (earthforce.org/green)
National Geographic Society (nationalgeographic.com)
National Recycling Coalition (nrc-recycle.org)
National Science Foundation (nsf.gov)
The Eisenhower National Clearinghouse (enc.org)
The Keystone Center (keystone.org)
The San Francisco Estuary Institute (sfei.org)
USFWS Education Resources (educators.fws.gov)
USGS Water Science for Schools (ga.water.usgs.gov/edu)
Water Quality Association (glossary) (wqa.org/glossary.cfm)

Environmental Organizations

Citizens for a Better Environment (cbew.org)
Clean Water Action (cleanwateraction.org)
Conservation Fund (conservationfund.or)
Earth First (earthfirst.org)

Earth Island Institute (earthisland.org)
Ecotrust (ecotrust.org)
Environmental Defense (environmentaldefense.org)
Freshwater Society (freshwater.org)
Friends of the Earth (foe.org)
Greenpeace (greenpeace.org)
Idaho Conservation League (wildidaho.org)
Montana Wilderness Association (wildmontana.org)
National Audubon Society (audubon.org)
National Wildlife Federation (nwf.org)
Natural Resources Defense Council (nrdc.org)
Northwest Environmental Advocates
    (northwestenvironmentaladvocates.org)
Oregon Natural Resources Council (onrc.org)
Sierra Club (sierraclub.org)
Southern Utah Wilderness Alliance (suwa.org)
The Nature Conservancy (nature.org)
The Wilderness Society (wilderness.org)
World Wildlife Fund (worldwildlife.org)

Erosion Control

International Erosion Control Association (ieca.org)

Federal Agencies

Environment Canada (ec.gc.ca)
National Oceanographic and Atmospheric Agency (noaa.gov)
National Research Council Canada (nrc.ca)
Natural Resources Conservation Service (nrcs.usda.gov)
United States Department of Interior (doi.gov)
United States Environmental Protection Agency (epa.gov)
United States Forest Services (fs.fed.us)
United States Geological Survey (usgs.gov)

Fisheries

Izaak Walton League of America (iwla.org)
Oregon Trout (oregontrout.org)

National Marine Fisheries Service (nmfs.noaa.gov)
Save our Wild Salmon (wildsalmon.org)
Trout Unlimited (tu.org)
United States Fish and Wildlife Service (fws.gov)

Groundwater

American Groundwater Trust (agwt.org)
The Groundwater Foundation (groundwater.org)
The Irrigation Association (irrigation.org)

Lakes/Ocean

American Society of Limnology and Oceanography (also.org)
Great Lakes Commission (glc.org)
Great Lakes Information Network (great-lakes.net)
Marine Advanced Technology Education Center
    (marinetech.org)
Oceanlink (marine education) (oceanlink.island.net)
SeaWeb (seaweb.org)

Reference Materials

Acorn Naturalists (acornnaturalists.com)
National Technical Information Service (ntis.gov)
United States Government Printing Office (access.gpo.gov)
United States Library of Congress (loc.gov)
Water Librarians Home Page (interleaves.org/~rteeter/
    waterlib.html)

Regulations

Code of Federal Regulations (access.gpo.gov/nara/cfr)

Research

American Geophysical Union (agu.org)
American Institute of Hydrology (aihydro.org)
American Water Resources Association (awra.org)
Center for Urban Forest Research (wcufre.ucdavis.edu)
Oregon State University CWEST (cwest.orst.edu)

National Institutes for Water Resources (wwri.nmsu.edu/niwr)
The Arizona Water Resource Research Center (ag.arizona.edu/azwater)

Rivers

American Rivers (amrivers.org)
European Rivers Network (rivernet.org)
International Rivers Network (irn.org)
Pacific Rivers Council (pacrivers.org)
Riverwatch Network (riverwatch.org)

State Agencies

Arizona State Dept. of Environmental Quality (adeq.state.az.us)
Florida Dept. of Environmental Protection (dep.state.fl.us)
Idaho Dept. of Environmental Quality (www2.state.id.us/deq)
Nevada Dept. of Natural Resources (state.nv.us/cnr)
New York State Dept. of Environmental Conservation (dec.state.ny.us)
Maine Dept. of Environmental Protection (state.me.us/dep)
Minnesota Dept. of Natural Resources (dnr.state.mn.us)
Oregon Dept. of Environmental Quality (deq.state.or.us)
Washington Dept. of Ecology (ecy.wa.gov)
Wisconsin Dept. of Natural Resources (dnr.state.wi.us)

Wastewater/Stormwater

National Small Flows Clearinghouse (estd.wvu.edu/nsfc)
Ohio's Water Professionals (ohiowater.org)
USEPA Office of Wastewater (epa.gov/owmitnet)
Water Environment Federation (wef.org)
Wastewater/water Products (wateronline.com)

Watersheds

Chesapeake Bay Foundation (cbf.org)
EPA Watershed Information Network (epa.gov/win)
Interagency Watershed Group (cleanwater.gov)

Oregon Watershed Enhancement Board (oweb.state.or.us)
The Oregon Plan for Salmon and Watersheds (oregon-plan.org)
The Trust for Public Land (tpl.org)
Water Connection (waterconnection.orst.edu)
Western Watershed Project (westernwatersheds.org)
William C. Kenney Watershed Protection Foundation
(kenneyfdn.org)

Web Directories

Environmental Web Directory (webdirectory.com/
water_resources)
General products (enature.com)

Wetlands

Ducks Unlimited (ducks.org)
Environmental Concern (wetland.org)
The Wetlands Conservancy (wetlandsconservancy.org)
USFWS National Wetlands Inventory Center
(wetlands.fws.gov)
US Society of Wetland Scientists (sws.org)
Wetlands International (wetlands.org)

Summary

Now that you have read this book, you are better informed about water quality and water pollution control. You learned about the connections between all parts of the water environment through the hydrologic cycle and about the natural and human factors affecting water quality. You learned the basics of water chemistry and microbiology. You read about the various sources of water pollution and how they can be prevented and controlled.

By reading this book, you found out about the rules and regulations that govern water quality protection in the United States. You learned about drinking water and the broad approach to protecting water quality by focusing on complete watersheds. Finally, you were given some ideas about how to get personally involved in protecting and maintaining clean water in your home and community.

This book provided you with a well-rounded introduction to the world of water quality and water pollution control. You may want to refer to it from time to time as you continue learning about and working for clean water.

Additional Reading

Grove, N., 1992. *Preserving Eden: The Nature Conservancy*. Nature Conservancy/Harry N. Abrams, Inc., New York, New York.

Harmonious Technologies, 1992. *Backyard Composting, Your Complete Guide to Recycling Yard Clippings*. Harmonius Press, Ojai, California.

Lines, L. (Ed.), 1973. *What We Save Now: An Audubon Primer of Defense*. National Audubon Society/Houghton Mifflen Company, Boston Massachusetts.

MacEachern, D., 1990. *Save Our Planet, 750 Everyday Ways You Can Help Clean Up the Earth*. Dell Publishing, New York, New York.

Mitchell, T., 1995. *Ecological Identity: Becoming a Reflective Environmentalist*. MIT Press, Cambridge, Massachusetts.

Palmer, T., 1986. *Endangered Rivers and the Conservation Movement*. University of California Press, Berkeley, California.

Seredich, J. (Ed.), 1991. *Your Resource Guide to Environmental Organizations*. Smiling Dolphin Press, Irvine, California.

Turner, T., 1991. *Sierra Club: 100 Years of Protecting Nature*. Sierra Club/Harry N. Abrams, Inc., New York, New York.

Glossary

Gunnison River, Colorado

Absorption: The process of removing pollutants by allowing them to contact and be taken into absorbent materials with an affinity for the pollutant.

Acid: A substance that, in aqueous solution, increases the hydrogen ion concentration, thereby lowering its pH. Acids have a pH lower than 7. Strong acids cause irritation and burning and can be toxic to aquatic organisms.

Activated carbon: A granular or powdered carbon material used as filter media to remove pollutants from water and wastewater. The carbon material is activated by heating it until it fractures, which creates many surfaces for pollutants to attach to.

Activated sludge: A biological treatment process where bacterial solids, called sludge, from a secondary clarifier are returned to the aeration basin to increase the active mass of microorganisms that are used for biodegrading waste materials. Activated sludge also refers to the mass of bacterial solids used in the treatment process.

Acute toxicity: The short-term effects of poisonous substances on the health of aquatic organisms, usually measured as mortality.

Adsorption: The process of removing pollutants by allowing them to adhere to the surface of materials they are attracted to.

Agronomic rate: The rate of application of wastewater or other fertilizers that allows crops to fully utilize the nutrients contained therein, preventing these nutrients—particularly nitrogen—from leaching through the soil and polluting groundwater.

Algae: A diverse group of single or multiple-celled, photosynthetic organisms. When present in large numbers, algae can cause large fluctuations in pH and dissolved oxygen in a water body and make it green and turbid.

Alkaline: A term used to describe water or other liquids that have a pH greater than 7. These substances are also called *basic substances.* Strongly alkaline substances can be harmful to aquatic organisms.

Ammonia: Form of inorganic nitrogen represented by the chemical symbol NH_3. Wastewater with high concentrations of ammonia can cause loss of oxygen, and toxicity, if discharged into a water body.

Anadromous fish: Species of fish, particularly Pacific salmon, trout, and char, that ascend rivers and streams from the ocean to spawn in fresh water.

Aquatic ecosystem: The environmental system of a stream, river, lake, estuary, or other water body and the habitat features and living organisms that exist in and around it.

Aquatic organisms: Organisms that live in or around water.

Aquifer: A soil formation that contains groundwater.

Assimilative capacity: The natural ability of a waterway to cleanse itself and assimilate waste materials through aeration, mixing, and biodegradation.

Back-washing: The process of washing a water or wastewater filter briskly with clean or recycled wash water circulated backwards through the filter, to dislodge and remove accumulated materials.

Bacteria: Simple, single-celled organisms that can be seen only under a microscope. Many different species of bacteria exist. Some are responsible for biodegradation and some can cause human disease.

Base: A substance that, in aqueous solution, has a pH greater than 7. These substances are capable of neutralizing acids. Household ammonia cleaner is an example of a simple base.

Base flow: The year-round flow in a river or stream that is provided by the continual inflow of groundwater and springs, and seeps into the river during both wet and dry seasons. Additional water added as a result of precipitation and runoff results in seasonally high flows.

Beneficial use: A recognized positive use made of the water in a river, lake or other water body. Beneficial uses include: domestic, industrial, and agricultural water supplies; fish and wildlife resources; power generation; contact recreation; aesthetic enjoyment; and others. Water quality standards are set to support the recognized beneficial uses of a water body.

Benthic organisms: Organisms that live on, in, or near the bottom of water bodies.

Best management practice (BMP): A management technique used to prevent or reduce water pollution, usually associated with nonpoint sources of pollution. Examples include: applying erosion control techniques, properly handling and storing materials, and using pesticides/herbicides/fertilizers properly, if at all. Normally, point sources of pollution are controlled by constructing wastewater treatment facilities. Nonpoint sources of pollution are controlled by applying best management practices.

Bioassay: A biological test used to determine the toxicity of the outflow from a wastewater treatment facility. These tests usually involve placing small organisms like minnows, water fleas, and algae in a sample of the effluent and evaluating toxicity by looking at the mortality, growth, and reproduction of the test organisms over time.

Biochemical oxygen demand (BOD): The amount of oxygen used by bacteria to biodegrade the organic material found in a sample of water or wastewater, usually measured over a five-day period. BOD is an indirect measure of the amount of organic material in the sample.

Biodegradation: The natural process whereby complex organic materials are broken down into simpler substances by microorganisms, particularly bacteria, as they utilize the organic material as a food source.

Biological treatment: A method of treating wastewater to remove organic material through the use of biodegradation; usually done in controlled settings, such as at municipal or industrial treatment plants, though it also occurs naturally.

Biomonitoring: A method of evaluating toxicity by placing aquatic organisms such as minnows, water fleas, and algae in samples of effluent. This term is also used more broadly to mean any evaluation that involves the use of aquatic organisms to determine environmental health.

Biosolids: The solids removed from wastewater treatment processes, consisting mainly of bacteria. The older term for these bacterial solids was sludge.

Bioswale: A ditch or swale planted with vegetation; used to remove pollutants from stormwater runoff through filtration, sorption, and, to a lesser degree, biodegradation.

Biota: All living organisms.

Brackish water: A mixture of fresh and salt water.

Buffer zones: A zone designed to separate two areas so that the impacts in one are not felt in the other. The area around a stream corridor that protects wetlands and open water from activities occurring on adjacent uplands.

Carbon: One of the most abundant elements on earth, represented by the chemical symbol C. All living organisms and other organic materials are made of carbon compounds.

Carcinogen: A substance that causes cancer.

Carcinogenic: Causing or contributing to the production of cancer.

Chemical oxygen demand (COD): The amount of oxygen used to chemically oxidize the organic material found in a water sample. COD is an indirect measure of the amount of organic material in the sample. The COD test is usually used to measure the organic contents of industrial wastewaters, whereas the biological oxygen demand (BOD) test is used for municipal wastewaters.

Chemical treatment: A method of treating water or wastewater by adding substances that cause chemical reactions. For example, a form of chemical treatment is adding bases to raise the pH of wastewaters containing dissolved metals so

that the metals form solids, or precipitates, that fall out of
the solution.

Chlorinated organics: Organic compounds that contain
chlorine and are formed when organic materials are subjected
to chlorine in processes such as bleaching paper and
disinfecting drinking water. Many of these compounds, also
called organochlorines, are toxic.

Chronic toxicity: The long-term effects of poisonous
substances on the health of aquatic organisms, determined by
measuring growth, reproduction, and mortality over time.

Clarification: The process of placing water and wastewater into
still basins, allowing suspended materials to settle from
solution—due to gravity—and then be removed.

Clarifier: A tank or basin in which water and wastewater are
placed to allow solids to settle from solution to prepare for
their removal.

Clean Water Act (CWA): The landmark piece of federal
environmental legislation protecting water quality in the
United States. It is based on a combination of federal water
laws dating back to the Rivers and Harbors Act of 1899. The
Clean Water Act contains many of the programs now used to
protect water quality, such as the National Pollutant
Discharge Elimination System program and the Water
Quality Standards program.

Code of Federal Regulations (CFRs): Documents published
by the Office of the Federal Register that contain all federal
regulations. The Code is divided into fifty Titles. Title 40
contains the federal regulations pertaining to protection of
the environment.

Combined sewer overflows: The overflow of sewage from
sewer pipes into stormwater pipes, which results in untreated
sewage being discharged directly into a water body. These
overflow events occur following rainstorms in communities
where the sewer and stormwater pipes are connected or
combined.

Comminutor: A mechanical device used at the entrance of wastewater treatment plants to cut or grind up solids in the incoming wastewater.

Compost: An organic fertilizer created by mixing biodegradable materials such as garbage, trash, lawn clippings, and prunings with soil and bacteria. The soil and bacteria decompose the biodegradable materials to create the fertilizer.

Comprehensive Environmental Response, Compensation and Liability Act (CERCLA): This act, which is also called the Superfund Act, was enacted by Congress in 1980 in response to the problems caused by abandoned hazardous waste disposal sites. It outlines a program for discovering abandoned or uncontrolled sites, evaluating the levels and types of contamination, and cleaning up the sites.

Constructed wetland: An upland area that has been turned into a manmade wetland by excavating and grading the area to form shallow ponds, planting aquatic vegetation, and adding water. Constructed wetlands are sometimes used for wastewater treatment by passing wastewater through them and allowing natural treatment processes to occur.

Desalination: The process of treating water to remove salts.

Disinfection: The process of treating water to kill harmful organisms such as bacteria and viruses, thereby protecting public health. Common methods of disinfection rely on the use of chlorine, ozone, or ultraviolet light.

Dissolved oxygen (DO): Oxygen gas dissolved in water, used by aquatic organisms for respiration in much the same way humans use oxygen in the air.

Drain field: The perforated underground pipes used to distribute wastewater flowing from a septic tank, and the soil area where these pipes are placed.

Drainage basin: All of the land area that surrounds and drains into a water body.

Ecosystem: All living organisms and nonliving features in an environmental community and the interactions between them.

Effluent: The outflow water from a wastewater treatment plant or septic tank. Also used to describe the outflow water from any part of the treatment process. For example, primary effluent is the outflow water from the primary treatment process.

Endangered Species Act (ESA): The federal law enacted in 1973 to protect animals, birds, fish, plants, and other living organisms from becoming extinct.

Environmental Assessment (EA): An environmental review document, usually brief, that describes the potential effects of federally sponsored activities on the environment, as required by the National Environmental Policy Act. Used to determine if a proposed federal action would have significant environmental effects.

Environmental Impact Statement (EIS): A detailed environmental review document that describes the potential effects on the environment of large or otherwise significant federally sponsored activities, as required by the National Environmental Policy Act.

Erosion: The wearing away of soil by the forces of water and wind. Although erosion occurs naturally, it can be increased by activities such as land clearing, road building, and timber harvesting, which may result in soil, debris, and other pollutants entering the water.

Eutrophic: A term applied to water bodies that have high concentrations of nutrients, resulting in the excessive growth of aquatic vegetation and algae and thereby reducing the clarity of the water and causing undesirable shifts in pH and dissolved oxygen.

Eutrophication: The process whereby clear, sterile, water bodies become nutrient-enriched with an abundance of algae and aquatic plants. This natural process can be accelerated by the activities of humans.

Evaporation: Loss of water into the atmosphere. Evaporation occurs at the surface of water bodies as water molecules change from liquid to gas as a result of temperature, wind, and humidity in the atmosphere.

Federal Emergency Management Agency (FEMA): The federal agency responsible for flood control, insurance, and disaster relief.

Fertilizer: Chemicals and other materials that contain the nutrients, such as nitrogen and phosphorus, required by plants and other living organisms for growth.

Filter strip: A strip of vegetation, such as grass, used to remove pollutants from stormwater by directing stormwater runoff through the vegetation.

Filtration: The process of removing solids by passing water through bars, screens, and filters that allow the water, but not the solids, to pass. Filtration also occurs in nature as water passes through vegetation or percolates through the soil.

Geographic information system (GIS): A computerized system for managing information by storing layers of data and relating it to geographic position. For example, a GIS could consist of a computerized map of a watershed showing the location and characteristics of different soil groups.

Giardia: A shortened name for the disease-causing microorganisms *Giardia lamblia*. These protozoans, found in water contaminated with wild-animal feces, cause a flu-like disease called Giardiasis.

Greenhouse effect: The insulating effect that atmospheric pollution, especially carbon dioxide, has on the earth, making it retain heat like a greenhouse.

Groundwater: Water found below the surface of the earth, stored in soil and rock formations.

Groundwater recharge zones: Areas where substantial quantities of surface water percolate into the earth and become part of the groundwater.

Hard water: Water that contains an abundance of calcium and magnesium, or other divalent cations, which are ions with two positive charges. Hard water causes the build up of mineral deposits on plumbing fixtures and in industrial boilers.

Headworks: The entrance portion of a wastewater treatment plant, usually consisting of screens and other equipment for removing solids and a device such as a flume for measuring flows.

Herbicides: Chemicals used to kill unwanted vegetation, such as weeds.

Hydric soils: Soils formed under conditions where they were saturated with water.

Hydrologic cycle: The unifying cycle in nature that connects all waters in the environment to one another, resulting from the movement of water due to evaporation, precipitation, and ground and surface water flow.

Hydrophytes: Vascular aquatic plants adapted for survival in saturated conditions.

Indicator organisms: Microorganisms used to indicate the presence of fecal contamination.

Infiltration gallery: A basin or similar area of rock or coarse soil used to place stormwater so that it will percolate through the material, remove some pollutants, and then infiltrate into the soil below.

Inorganic: All materials that are not made from plants, animals, or synthetic carbon compounds. Rocks, minerals, and metals are inorganic compounds.

Lake overturn: The movement of water in deep lakes, initiated by changes in the season that cause the temperature and density of the water at the top and bottom of the lake to reverse. It occurs as the water at the surface of the lake becomes colder and heavier than the water below it. Lake overturn, which may occur in the fall and spring, causes a lake to mix and water chemistry in the lake to become more uniform.

Large Quantity Generators: Those facilities generating more than 2.2 pounds of acute hazardous waste or more than 2,200 pounds of any hazardous waste, as defined by the Resource Conservation and Recovery Act.

Leachate: The wastewater formed when rain or other surface water moves through the waste materials in a landfill. This high-strength wastewater normally has high concentrations of dissolved metals and other harmful pollutants.

Leaf-compost filter: A special type of filter made of composted leaves and used to remove pollutants from stormwater.

Limiting nutrient: The nutrient in shortest supply that limits the growth of aquatic organisms like algae. Phosphorus is the limiting nutrient in most aquatic systems, although nitrogen and other micronutrients can also be limiting.

Load allocation: The maximum load of a pollutant allocated to nonpoint and background sources of pollution discharging into a waterway. Load allocations are used to limit pollutant loadings into water quality limited waterways so that water quality standards can be achieved.

Maximum Contaminant Level (MCL): The maximum concentration of a contaminant allowed in drinking water under the Safe Drinking Water Act to protect public health.

Mesotrophic: The middle trophic state of a water body; between oligotrophic and eutrophic, characterized by a moderate concentration of nutrients supporting the growth of some aquatic organisms.

Methemoglobinemia: A disease in infants caused by ingestion of water containing nitrogen in the form of nitrate (NO_3). Nitrate prevents oxygen from circulating properly in the blood stream, resulting in suffocation in severe cases. This ailment is also called blue baby disease.

Milligrams per liter (mg/L): A term used to describe the concentration of a substance in water by expressing the weight or mass, in milligrams, of the substance found in one liter of water.

Mixing zone: That portion of a receiving stream below a permitted discharge where effluent is allowed to mix with ambient water prior to meeting water quality standards. Mixing zones are defined in NPDES permits. A typical mixing zone could be defined as that area within a fifty-foot radius from the point of discharge.

Monitoring: Taking samples or measurements to determine the health of an ecosystem, such as a river, lake, or watershed. Water quality monitoring involves collecting water samples and evaluating them in the laboratory to determine the chemical makeup of the waters they represent. Effluent monitoring, required by discharge permits, involves taking samples of effluent and having them analyzed for the parameters listed in the permit to determine compliance.

Morphology: Characteristics of water bodies, such as their depth, width, area, and shape.

National Environmental Policy Act (NEPA): The law passed by Congress in 1969, directed at evaluating the environmental impacts of all federally sponsored activities. Environmental review documents prepared under this law take the form of either an Environmental Assessment (EA) or an Environmental Impact Statement (EIS).

National Pollution Discharge Elimination System (NPDES) Permit: A permit issued to municipalities and industries that allows them to discharge treated wastewater into waters of the United States. These permits, authorized under Section 402 of the Clean Water Act, specify the degree of treatment needed prior to discharge, as well as the type and frequency of testing to be conducted on the effluent. NPDES permits also apply to some stormwater discharges.

National Priority List (NPL): A list of more than 1,200 abandoned or uncontrolled hazardous waste disposal sites in the United States, identified under the Comprehensive Environmental Response, Compensation and Liability Act (CERCLA), or Superfund.

Neutralization: The process of adjusting the pH of an aqueous solution so that it reaches a neutral value of 7.0, which is neither acidic nor basic. Acidic solutions are neutralized by adding bases to them and basic solutions are neutralized by adding acids.

Nitrate: A form of inorganic nitrogen resulting from the oxidation of ammonia, identified by the chemical symbol NO_3. Nitrate may cause respiratory problems in infants if they drink water or formula with nitrate concentrations greater than ten milligrams per liter. Nitrate has also been found to cause cancer in laboratory animals.

Nitrogen: An abundant element in the environment used to form amino acids, which are the building blocks of proteins. Nitrogen, identified by the chemical symbol N, is one of the essential nutrients for plant and animal growth. Its various forms—organic nitrogen, ammonia, and nitrate—play important roles in water quality and water pollution control.

Nitrosamines: Compounds derived from nitrate that pose a potential cancer risk. Nitrosamines can be formed in the body from nitrate found in food or water.

Nonpoint source: A source of pollution that comes from a broad area rather than a single point of origin. Stormwater runoff is an example of a type of nonpoint source pollution. According to the EPA, nonpoint sources of pollution are the leading cause of water pollution in the United States today.

Nontoxic: Not poisonous or harmful to living organisms.

Nutrients: Elements such as carbon, nitrogen, and phosphorus that are necessary for the growth of all living things. Nutrients are the chemical building blocks of all plants, animals, and humans.

Oil/water separator: A device used to separate oil from water, based on the principle that oil is lighter than water, so it floats.

Oligotrophic: The youngest trophic state of a water body, characterized by almost sterile water that contains few nutrients or organisms and abundant dissolved oxygen.

Organic: Made from plants or animals or created from carbon compounds in a laboratory.

Organism: Any living thing.

Parts per million (ppm): A unit of measurement used to describe the concentration of substances in water. Equivalent to milligrams per liter (mg/L). For example, a concentration of 10 ppm of nitrogen—meaning that ten parts of nitrogen exist for every million parts of water—is equivalent to 10 mg/L.

Pass-through pollutants: Substances that pass through conventional wastewater treatment processes without being treated or removed. Special pretreatment processes must remove these pollutants before discharging industrial wastewaters into municipal wastewater treatment plants.

Pathogenic: Disease-causing. Microorganisms that cause diseases in humans, such as some species of bacteria and viruses, are called pathogenic organisms.

Pesticides: Chemicals used to kill pests such as insects.

pH: A chemical term used to describe the acidic or alkaline nature of a liquid due to the concentration of hydrogen ions. Technically defined as the negative logarithm of the molar concentration, or activity, of hydrogen ions.

Phosphorus: One of the essential nutrients for biological growth that can contribute to the eutrophication of lakes and other water bodies. Increased phosphorus levels result from the discharge of phosphorus-containing materials such as fertilizers and detergents into surface waters. Phosphorus is represented by the chemical symbol P.

Physical treatment: A method of treating wastewater to remove pollutants by using screens, filters, or other devices that physically separate the pollutants from the water, or by allowing pollutants to settle from solution due to gravity.

Point source: A source of pollution that originates from a single point, like the discharge end of a pipe. Municipal and industrial discharges are point sources of pollution.

Precipitates: Solid materials that form when dissolved substances combine as a result of chemical reactions. For example, metal solids or precipitates form when an alkaline substance is added to a solution containing dissolved metals, thereby increasing the solution's pH.

Primary clarifier: The first large tank or basin at a treatment plant where wastewater is placed to allow the heavier solids to settle from solution.

Primary containment: The first means of holding oil and other potentially harmful substances, usually a metal tank.

Primary Drinking Water Standards: Standards established under the Safe Drinking Water Act to protect public health, consisting of the maximum concentration of specific contaminants allowable in drinking water. Also called Maximum Contaminant Levels or MCLs.

Primary treatment: The first level of wastewater treatment provided at municipal plants using the physical processes of screening and sedimentation. The effluent from primary treatment is not of satisfactory quality to be discharged into a receiving stream without causing pollution. Primary effluent must receive further biological treatment—referred to as secondary treatment—and disinfection prior to discharge.

Protozoa: A type of multiple-celled microorganism commonly found in water, soil, and sewage. Most protozoa found in the environment feed on bacteria. Some harmful species of protozoa cause human diseases, such as amebic dysentery and Giardiasis.

Recycling: Conserving natural resources by using items more than once in the same or alternate forms.

Reducing: Conserving natural resources by using less of an item or creating less waste.

Resource Conservation and Recovery Act (RCRA): A law enacted by the United States Congress in 1976 for the primary purpose of ensuring that hazardous wastes are managed properly from the time they are generated until they are ultimately disposed of or destroyed.

Reusing: Conserving natural resources by using the same item more than once.

Safe Drinking Water Act (SDWA): An act signed into law by Congress in 1974 to protect public health by keeping drinking water free from contamination. It is the key piece of legislation protecting drinking water in the United States, and establishes both Primary and Secondary Drinking Water Standards.

Sample blanks: Samples of distilled water that are tested along with effluent or ambient water samples to evaluate laboratory procedures and ensure quality results.

Secondary clarifier: A large basin used to settle out biological solids at municipal wastewater treatment plants. Biological solids from the plant's aeration basin or trickling filter flow into the secondary clarifier, where they settle to the bottom and are later removed.

Secondary containment: A second means of holding or containing spilled liquids in case the first means of containment fails. As a general rule, metal tanks filled with petroleum products are placed in concrete bunkers so that if one of the tanks break, the gasoline or oil will still be contained. The tanks provide primary containment. The concrete bunkers provide secondary containment.

Secondary Drinking Water Standards: Standards established by the Safe Drinking Water Act to protect the aesthetic qualities of drinking water, such as its appearance, taste, and odor.

Secondary treatment: The process of removing pollutants, particularly organic materials, from municipal wastewater through controlled biodegradation. Bacteria use the organic material in the wastewater as food and energy for growth and reproduction, converting the organic material into new bacterial cells that are later removed.

Sedimentation: The process of suspended materials falling, or settling, out of the water as a result of gravity; one of the fundamental processes that remove solid pollutants from water in nature and at water and wastewater treatment plants.

Septic tank: A 500 or 1,000 gallon tank made of concrete or fiberglass used to hold sewage solids, usually from individual homes. This tank and the associated pipeline and drain field, used to distribute the liquid effluent from the tank into the soil, make up a septic system.

Small Quantity Generators: Facilities generating more than 220 pounds but less than 2,200 pounds of hazardous waste, as defined under the Resource Conservation and Recovery Act (RCRA).

Solvent: A liquid that is capable of dissolving or dispersing one or more other substances.

Sorption: The combined processes of adsorption and absorption, used to remove pollutants from water and wastewater.

Spill Prevention Control and Countermeasure (SPCC) Plan: A plan required by Section 311 of the Clean Water Act for sites where petroleum products could contaminate waters if spilled. These documents include procedures to prevent spills from occurring and to respond effectively if they do.

Stormwater residuals: Solids and other pollutants that are left in stormwater treatment facilities following treatment.

Stormwater runoff: Water that runs off of the surface of the land and other surfaces after a rainstorm or snowstorm.

Stratification: The process whereby separate layers of water develop in a water body, each with different physical and chemical characteristics. For instance, thermal stratification may occur in a deep lake if the water at the top and bottom develop distinctly different temperatures.

Superfund: The common term applied to the Comprehensive Environmental Response, Compensation and Liability Act.

Surface water: All water that is on the surface of the earth, including streams, rivers, lakes, and the ocean. This term is frequently used to differentiate water on the surface from groundwater, which resides below the surface of the earth.

Tertiary treatment: A form of advanced wastewater treatment that goes beyond conventional primary and secondary treatment to further treat and "polish" effluent for special uses. May involve the use of special filters or chemical processes.

Thermal pollution: Any form of pollution that causes an increase in the temperature of the water.

Total dissolved solids (TDS): A measure of the solids in a water sample that are so small they are essentially dissolved and can pass through a fine paper filter, expressed in units of concentration such as milligrams per liter.

Total maximum daily load (TMDL): The quantity of material allowed to be discharged into a waterway from all recognized sources while still maintaining applicable water quality standards. TMDLs are calculated when a water body does not meet water quality standards to determine the maximum allowable load prior to implementing pollution control alternatives. The TMDL is the sum of the individual waste load allocations from point sources and load allocations from nonpoint sources and background, plus the amount set aside for reserve.

Total solids (TS): A measure of all the solids in a water sample, including those that are suspended and dissolved, expressed in units of concentration such as milligrams per liter.

Total suspended solids (TSS): A measure of the solids in a water sample that are too large to pass through a fine paper filter, expressed in units of concentration such as milligrams per liter.

Toxic: Harmful to living organisms. Poisonous.

Trybutilin (TBT): An additive used in marine paints to prevent barnacles from growing on the hulls of boats. TBT causes chronic toxicity. Aquatic organisms exposed to it, such as shellfish, become deformed.

Trihalomethanes (THMs): Potentially cancer-causing substances that form when chlorine is added to water containing organic materials.

Viruses: Tiny microorganisms—approximately ten to one hundred times smaller than bacteria—that cannot be seen without a special type of microscope. Viruses are parasites; they cannot live outside the cell of another organism, which is called the host. Viruses are responsible for human diseases such as smallpox, infectious hepatitis, influenza, and poliomyelitis.

Waste load allocation: The maximum load of a pollutant each point source discharger is allowed to release into a particular waterway. Waste load allocations are used to limit point source discharges into water quality limited waterways so that water quality standards can be achieved.

Water Quality Certification: The process of obtaining approval from state environmental agencies to conduct activities that may affect water quality and are subject to federal permits. Water Quality Certification is required by Section 401 of the Clean Water Act for activities such as building dams (which requires a federal permit from the

Federal Energy Regulatory Commission) and filling or removing material from wetlands (which requires a federal permit from the Army Corps of Engineers).

Water quality limited: Water bodies that do not meet water quality standards for parameters such as temperature, pH, or dissolved oxygen.

Water quality standards: The values or statements used to define the acceptable characteristics of water bodies, mandated by Section 303 of the Clean Water Act. Water quality standards for characteristics such as pH, dissolved oxygen, and temperature are developed by individual states and submitted to the Environmental Protection Agency for approval.

Water softening: The process of removing calcium, magnesium, and other divalent cations, elements containing two positive charges. In the simplest type of water softening, calcium and magnesium are removed by passing water through a container of salt. The calcium and magnesium trade places with the sodium in the salt through a process called ion exchange, removing them from the water.

Watershed: A portion of land that all drains into the same water body; also called a drainage basin. Precipitation falling into a watershed flows downhill from the mountaintops and ridges, which are the boundaries of the watershed, to the streams and lowlands below. Watersheds are made up of all the land forms, vegetation, organisms, and waters in the basin.

Wetland: An area that is saturated with water at a sufficient frequency to support the growth of vegetation adapted for life in saturated soil conditions. Many different type of wetlands exist and they go by many different names such as bogs, swamps, marshes, and wet meadows.

Zone of initial dilution (ZID): That area within a permitted mixing zone where standards for acute toxicity may be waived.

Index

Page numbers in italics refer to illustrations and photographs.